Sperm-Mediated Gene Transfer: Concepts and Controversies

Editor

Kevin R. Smith

Abertay University
UK

CONTENTS

FOREWORD

One hundred years ago, 1911, Peyton Rous discovered a "filterable agent", Rous Sarcoma Virus, to be the cause of wing-web tumors in chickens on a local farm in New York State. That viruses could cause cancer was unaccepted by the scientific establishment, largely because the epidemiology of cancer was inconsistent with that of known viruses, such as influenza. It was not until the mid 1930's, when Richard Shope demonstrated that the warts on the plantar surface of rabbit feet were caused by a virus, and together with Peyton Rous, demonstrated that the papilloma virus could also induce malignant tumors, that there was an initial acceptance of a virus etiology of cancer. There ensued a global hunt for viruses that cause cancer in mammals, a quest that has been largely unsuccessful. There remains the biological puzzle: why would viruses cause leukemias and sarcomas in chickens, but not in humans?

Fifty years after its discovery, Howard Temin reported another characteristic of Rous Sarcoma Virus that would rock the scientific community: its replication required an enzymatic step that made a DNA copy of its RNA genome. The concept of DNA to RNA to Protein as a "central dogma of biology" had been widely accepted, and the notion that information could flow from RNA to DNA was unaccepted by the scientific establishment, and not in keeping with Mendelian genetics. Like Peyton Rous, Howard Temin's work was discredited and difficult to get published.

This one avian virus had rocked several basic tenets of biomedicine: that cancer could be caused by an infectious agent, that information could flow in both directions between DNA and RNA, that there were major species-specific differences in disease causation, and, finally, that the oncogenic nature of the virus was due to its incorporation and slight mutation of a normal gene, *Src*, encoding a protein kinase important to cell division.

In 1971, sixty years after the discovery of the virus, Temin hypothesized that information flow from messenger RNA to DNA provided a mechanism for short-term amplification of gene sequences, perhaps fundamentally important to embryonic development; this hypothesis remains largely untested. The same year, Ben Brackett, in the laboratory of Hilary Kiprowski, demonstrated the uptake of Simian Virus 40 DNA into rabbit sperm heads, and the delivery to rabbit ova at fertilization.

In 1989, apparently unaware of the 1971 report by Brackett, Spadafora and colleagues described the uptake of exogenous DNA by mouse sperm, and the generation of transgenic pups; and Arezzo reported DNA uptake by three species of sea urchins and the generation of transgenic sea urchin embryos expressing the reporter gene contained in the exogenous DNA. At least two large U.S. laboratories rushed to repeat the mouse work in an effort to more easily create transgenic mice, but because their efforts were unsuccessful and widely reported, the scientific community did not embrace this new phenomenon of "sperm mediated gene transfer," SMGT.

Now, however, forty years after the original observation in rabbit ova, SMGT has become an increasingly important tool, as reported in this eBook, whose chapters are rich with new, species-specific approaches to SMGT, and new animal husbandry and biomedical applications. And to add to the consternation of the scientific community, there is now ample evidence for an endogenous reverse transcriptase in sperm that plays a fundamentally important role in SMGT, including genetic information reverse transcribed from exogenous RNAs taken up by sperm, "sperm mediated reverse gene transfer," SMRGT.

The problem for the scientific establishment is "What does this mean?" Is there another form of inheritance, distinct from Mendelian genetics? There is evidence to support this notion. And was Temin correct, that reverse transcription plays a fundamentally important role in early embryonic development, perhaps beginning with fertilization? There is also mounting evidence to support this notion. Could an individual's disease propensity be established at fertilization by exogenous RNAs or DNAs encountered by the sperm in its journey to the egg?

What would be the advantage to an organism incorporating exogenous RNA or DNA, perhaps pathogenic, into a fertilized egg? Although HIV appears to be a new human pathogen, it is now clear that human cells have an innate defense against reverse transcription, the APOBEC3G family of cytidine deaminases. Hence, robust defense mechanisms to protect the integrity of the genome itself are probably operational, especially in the egg, throughout nature. But replicating extra-chromosomal RNAs and DNAs may also be a routine cell function, challenging again the central dogma of DNA to RNA to protein.

Sperm Mediated Gene Transfer: Concepts and Controversies includes essential background information for newcomers to the field, summarizes the work accomplished, and heralds the work to be done -- to begin to understand and fully employ the transfer of exogenous genetic information at fertilization. Peyton Rous and Howard Temin would be fascinated by the following chapters.

Ann A. Kiessling
Harvard Medical School
Bedford Research Foundation
USA

PREFACE

As a young researcher in an established transgenic laboratory at Edinburgh University in 1989, I was struck by the effects of the publication in the journal *Cell* of a paper reporting that mammalian sperm could readily act as vectors for foreign DNA. Almost overnight, attention in our laboratory switched from established methods of transgenesis towards this exciting new prospect, and immediate efforts were made to replicate the *Cell* work. The talk at the time was that all the paraphernalia, training and expense of pronuclear microinjection (the standard approach) would be rendered redundant by this disruptive new technology, known as sperm-mediated gene transfer (SMGT).

In the event, the original *Cell* work could not be replicated, either in our laboratory or in a number of laboratories worldwide that were similarly attempting to make SMGT work. This failure was experienced by many in the field of animal transgenesis as something of a body-blow. Researchers were genuinely very disappointed that SMGT manifestly did not deliver on its initial promise. A backlash followed, with a substantial amount of skepicism being directed towards both the *Cell* paper and also towards the fundamental biological notion that sperm could ever be expected to be able to act as transgene vectors. Almost as rapidly as interest in SMGT had exloded, most researchers abandoned their efforts towards getting sperm to carry foreign DNA.

Nevertheless, a small, disparate array of transgenic scientists continued to work with sperm in the context of SMGT, either in the hope of establishing SMGT as a viable method for producing transgenic animals, or to address fundamental questions concerning interactions between sperm and exogenous nucleic acid molecules and the parameters influencing such activity. The outcome from such work has been as varied as the research purposes involved. Since the late 1980s, several significant papers describing sperm-DNA interactions have been published. For example, we are now much more knowledgeable about the fate of nucleic acids taken up by sperm; it has become clear that certain forms of augmentation, for instance the combination of DNA incubation with intracytoplasmic sperm injection (ICSI), permits a reliably high level of exogene uptake; and transgenic animal generation has been reported *via* highly novel forms of SMGT, including *in vivo* transgene injections into the male reproductive tract and the use of nanotechnology to deliver transgenes into sperm.

Yet many unanswered questions remain, and certain tensions exist within the body of empirical data and theory associated with SMGT. To what extent may horizontal (sperm-mediated) gene transfer occur in nature? To what extent do the models proposed to describe exogene uptake & integration reflect reality, and to what extent may such mechanisms generalise to all animals? Why have a few groups continued to report extremely impressive results for the generation of transgenic animals using the original unaugmented ('autouptake') SMGT methodology, where such results could not be replicated by other groups? And how do such positive reports fit with current theory (itself based on empirical research) that suggests sperm are unlikely to permanently harbour integrated transgenes following autouptake?

While such questions and controversies have yet to be fully answered or resolved, it can be stated with confidence that continued research in the context of SMGT has, at the very least, significantly expanded our understanding of sperm cellular and molecular biology. In addition, ongoing research into SMGT has kept alive the tantalising possibility that sperm have the potential to be routinely used for important genetic modification applications, including the generation of transgenic disease models, improved agricultural strains, transgenic bioreactors, xenotransplantation technology and perhaps even human gene therapy.

The contributors to this book elucidate a broad range of theoretical and empirical aspects of SMGT. The overall result is the construction of an intriguing picture of the diversity of potential applications and implications arising from the possible use of sperm as genetic vectors. A wide array of animal types is covered, ranging from invertebrates to large farm animals, and the range of SMGT augmentation methods described is similarly extensive. And the aforementioned controversial aspects of SMGT are evident in these writings.

This book, with its expert contributions, should allow the reader an unparalleled depth of insight into the concepts and controversies of SMGT. It remains to be seen whether the SMGT revolution promised by the publication of the 1989 *Cell* paper will ever materialise. But the reader will be left in no doubt as to the importance of SMGT to modern bioscience.

Kevin R. Smith
Abertay University
UK

List of Contributors

Pablo Bermejo-Álvarez Department of Animal Reproduction, INIA, Ctra de la Coruña KM 5.9, Madrid 28040, Spain

Sebastian Canovas University of Murcia, Spain

Tiago Collares Biotechnology Centre, Federal University of Pelotas, Brazil

Odir A. Dellagostin Biotechnology Centre, Federal University of Pelotas, Brazil

João Carlos Deschamps Biotechnology Centre, Federal University of Pelotas, Brazil

Vinicius Farias Campos Biotechnology Centre, Federal University of Pelotas, Brazil

Raúl Fernández-González Department of Animal Reproduction, INIA, Ctra de la Coruña KM 5.9, Madrid 28040, Spain

Celia Frutos Department of Animal Reproduction, INIA, Ctra de la Coruña KM 5.9, Madrid 28040, Spain

Joaquin Gadea University of Murcia, Spain

Francisco Alberto Garcia-Vazquez University of Murcia, Spain

Nasser Ghanem Cairo University, Egypt

Michael Gurevich Tel HaShomer Hospital, Israel

Alfonso Gutiérrez-Adán Department of Animal Reproduction, INIA, Ctra de la Coruña KM 5.9, Madrid 28040, Spain

Eliane Harel-Markowitz Kimron Veterinary Institute, Israel

Michael Hölker University of Bonn, Germany

Ann Kiessling Harvard Medical School, Bedford Research Foundation, USA

Ricardo Laguna Department of Animal Reproduction, INIA, Ctra de la Coruña KM 5.9, Madrid 28040, Spain

Carlos Frederico Ceccon Lanes University of Nordland, Norway

Luis Fernando Marins Universidade Federal do Rio Grande (FURG), Brazil

Alberto Miranda Department of Animal Reproduction, INIA, Ctra de la Coruña KM 5.9, Madrid 28040, Spain

John Parrington University of Oxford, United Kingdom

Miriam Pérez-Crespo	Department of Animal Reproduction, INIA, Ctra de la Coruña KM 5.9, Madrid 28040, Spain
Karl Schellander	University of Bonn, Germany
Ilaria Sciamanna	Italian National Health Institute, Viale Regina Elena 299, 00161 Rome, Italy
Fabiana Kömmling Seixas	Biotechnology Centre, Federal University of Pelotas, Brazil
Mordechai Shemesh	Kimron Veterinary Institute, Israel
Laurence Shore	Kimron Veterinary Institute, Israel
Kevin R. Smith	Abertay University, United Kingdom
Corrado Spadafora	Italian National Health Institute, Viale Regina Elena 299, 00161 Rome, Italy
Yehuda Stram	Kimron Veterinary Institute, Israel
Dawit Tesfaye	University of Bonn, Germany
Wei Shen	Qingdao Agricultural University, China
Xiaofeng Sun	Qingdao Agricultural University, China
Yidong Niu	Peking University People's Hospital, China

2

CHAPTER 1

Sperm-Mediated Gene Transfer: History and Background

Corrado Spadafora[*]

Italian National Health Institute, Viale Regina Elena 299, 00161 Rome, Italy

Abstract: The concept of Sperm-Mediated Gene Transfer (SMGT) describes the ability of spermatozoa to deliver to embryos not only their own genome during fertilization, but also foreign genetic information with which they may come in contact. The concept was established in 1989, when it was first shown that exogenous DNA incubated with mouse spermatozoa can be detected in tissues of born offspring. That initial finding has promoted a wealth of studies, progressively extended to a variety of species and still ongoing, with the dual aim to a] clarify the basic molecular mechanisms underlying SMGT on the one hand, and, b] on the other hand, develop biotechnology applications to generate genetically modified animals. Progress in any scientific field is often achieved in a non-linear manner and re-examining its historical unfolding serves to formalise conceptual advance as well as any remaining gaps. In this chapter I recapitulate the research progress and conceptual evolution within the SMGT field since its initial discovery. I focus on basic studies and particularly discuss findings that I consider as the distinctive milestones of the field. This historical journey spans more than twenty years, from the earliest observations that first suggested the idea that sperm cells are permeable to exogenous molecules to the recent finding that a reverse transcriptase (RT)-dependent mechanism is active in sperm cells and can regarded as a continuous source of novel genetic traits.

Keywords: Sperm chromatin, Nuclease sensitivity, "Active" sperm chromatin, Sperm/DNA binding, Internalization of exogenous DNA into sperm nucleus, SMGT, Transgenesis, Integration in sperm genome, Extrachromosomal sequences, Reverse Transcriptase, Non-mendelian inheritance.

FOUNDING GENETIC TRANSFORMATION OF SPERM

The pioneering finding of Avery, Macleod and McCarty in 1944 [1], which identified the DNA and not proteins as the repository of genetic information, and the 1953 groundbreaking article of Watson and Crick [2], proposing the double-helix DNA model, are among the historical achievements that have exerted the greatest impact on biological thought, influenced the direction of modern biology and provided a foundation for the current explosion of genomic studies. By establishing that DNA is indeed responsible for the transmission of all genetic traits and the shrine of the genetic identity of every cell, those early findings have disclosed for the first time the realistic possibility that both genetic traits and cell identity could be modified by transferring DNA fragments of exogenous origin inside cells.

The experimental genetic modification of cell traits using DNA from exogenous sources reflects a phenomenon currently occurring in nature called "horizontal (or lateral) gene transfer" (HGT) which consists in the passage of genetic information across mating barriers between distantly related organisms such as prokaryotes and lower eukaryotes (for a review see [3]) but recently has also been reported in animals [4].

The last few decades have witnessed a burst of efforts aiming at transferring foreign DNA in a long list of recipient living cells and organisms, including prokaryotic and eukaryotic cells and tissues, as well as oocytes in order to establish genetically transformed new strains, in a variety of different organisms. In parallel, a considerable amount of effort has been invested to develop appropriate technologies to facilitate the delivery of DNA into host cells, *e.g.* improved transfection protocols, delivery systems, biological vectors and microinjection devices. All these efforts have promoted both basic research studies and applied technology and have supported the emergence of new highly specialized fields, such as cell therapy, gene therapy and animal transgenesis. Animal transgenesis, in particular, has attracted enormous attention,

*Address correspondence to Corrado Spadafora: Italian National Health Institute, Viale Regina Elena 299, 00161 Rome, Italy; Tel: 39 (0)6 49903117; E-mail: corrado.spadafora@iss.it

efforts and resources, due to its powerful promise that manipulating animal genomes could help the understanding and therapy of many human genetic conditions and diseases [5, 6], as well as contributing a valuable biotechnology tool of socio-economical impact for those species of agricultural/zootechnological interest [7, 8]. In the last decades, virtually all experimentally relevant types of cells, both somatic and from the germ lineage, have been targeted and successfully transformed with either exogenous DNA or RNA molecules.

In the crowded variety of cell types that have proved amenable to transfection, the remarkable exception of mature spermatoza stands out. Sperm cells were not tested as possible recipient cells in gene delivery experiments, nor were they even theoretically thought of as reasonable candidates for transfection with foreign DNA. This preconceived exclusion has long been unquestioned by the scientific community and reflects a spontaneous consensus on what we know now to be a prejudice, but then perceived as a self-explaining evidence not needing experimental proof, namely that spermatozoa are highly compact cells, metabolically inert and virtually impenetrable to exogenous molecules. This view or prejudice was based on the unusual morphology of spermatozoa, which sharply diverges from that of all other cells; these morphological features include a long tubulin-based flagellum providing the motility apparatus, an almost total absence of cytoplasm, and a highly condensed chromatin organized in a nucleo-protamine complex tightly packed in the sperm head. These distinctive structural features did not seem to be compatible with complex biochemical functions or cellular activities, leading to the idea that spermatozoa are metabolically inert cells, whose peculiar and exclusive function is the transmission of the paternal genome during fertilization.

My personal view of sperm cells began to change in the late 70's, while studying the organization of sea urchin sperm chromatin using micrococal and pancreatic (DNaseI) nuclease digestion [9, 10]. At the time, chromatin digestion with nucleases was a popular and highly informative tool to study the higher-order organisation of the genome based on the discovery that the enzymes cleave preferentially the "accessible" chromosomal DNA that is not protected by nucleosomes, whereas the nucleosomal DNA, wrapped around the histone octamer core, remains protected from cleavage, thus generating a typical multimeric nucleosomal ladder upon DNA fractionation by gel eletrophoresis [11]. In this framework, we were seeking to characterise the sea urchin sperm chromatin and particularly the arrangement of the "accessible" linker DNA placed between adjacent nucleosomes [12, 13]. The surprising finding that puzzled me was that chromatin cleavage was produced equally well either using classical protocols based on nuclei isolation, or by simply adding the nuclease to buffered sea water in which intact spermatozoa were swimming: under the latter conditions, spermatozoa quickly stopped moving and, after DNA extraction, produced the typical nucleosomal ladder [14]. These observations implied that, in the absence of any treatment, the nucleases added to the medium were capable to cross both the plasma and the nuclear membranes and gain access to the chromosomal DNA of living spermatozoa. To my knowledge, this was the first evidence for the existence of a direct passage between the external environment and the inside of the sperm nucleus.

That finding shattered my firm conviction that sperms are impenetrable. Gradually, the concept emerged that, if protein molecules such as nucleases were able to penetrate spontaneously the sperm nucleus, then exogenous DNA molecules might also do the same in principle. The idea, initially conceived as a vague and improbable suggestion, gradually took the shape of a consistent possibility and was eventually formalised into an experimentally testable hypothesis. If amenable to experimental demonstration, the ability of spermatozoa to take up exogenous DNA would impose a radical reappraisal of their biological role, not only as vectors of their own genome, but also as potential recipients and carriers of genetic information of exogenous origin. In addition, if foreign sequences could be demonstrated to be delivered from sperm cells to oocytes at fertilization and propagated in the developing embryos, this finding may have far-reaching implications for the genetic identity of the progeny.

What follows is a personal account of the studies that unfolded from the initial observation of sperm permeability, culminating with the formalisation of SMGT, of the controversy that followed and of the knowledge that has been gained from efforts to develop the field. The field of SMGT has become established through various phases, and recapitulating this history is useful to rationalise what we now understand and can currently master as well as delineate the questions that are still unsolved. Most basic studies aiming to clarify the mechanism of the interaction between sperm cells and DNA have been carried out in my laboratory in the last twenty years. Over time, an unanticipated complexity has emerged; efforts to understand this complexity have recently led to identify a novel reverse transcriptase-dependent mechanism that provides spermatozoa with the potential to act as autonomous sources of novel genetic traits, unlinked to chromosomal genes.

SPERM MEDIATED GENE TRANSFER (SMGT), THE FIRST PHASE

In the early '80s, initial attempts to investigate whether spermatozoa could act as vectors of foreign DNA molecules in my laboratory made use of *X. laevis* spermatozoa and P^{32}-labelled plasmid DNA. That work did not provide significant clues on the nature of the binding of DNA to sperm cells, but helped to pinpoint a number of key parameters: it suggested that labelled DNA fragments could indeed bind to spermatozoa - a proportion of the labelled DNA remaining tightly sperm-bound even after multiple washing rounds - and that some of the foreign molecules could be delivered to oocytes at fertilization. In addition, some of the Southern blotting data at the time could be taken to indicate that some DNA molecules were propagated in very early embryos. On the whole, the results from these experiments were poorly consistent and not always unambiguous, but in retrospect they did provide a valuable core of preliminary data that confirmed aspects of the original hypothesis and fuelled our further investigation into this phenomenon. We now know that the weakness of the results was caused by the presence, in *X. laevis* semen, of factor(s) placed on the sperm cell surface that inhibit DNA binding. We were unaware of the existence of such inhibitory factors at the time that these experiments were being carried out. The work with *X. laevis* did not seem solid enough to support the conclusion that sperm cells are indeed vectors of foreign genetic information and was therefore never submitted to be published.

A turning point came with the decision to abandon the *X. laevis* system and to continue the study in the murine system. A well-known protocol for *in vitro* fertilization (IVF) [15] was adapted to this aim. After the long desultory experience with the *X. laevis* system, it was clear to me that undisputable, compelling evidence that spermatozoa can be vectors of exogenous DNA could only come from the generation of transgenic mice, or at least from the identification of the foreign sequences originally incubated with sperm cells in tissues of the offspring. Mouse IVF is a more complicated and demanding experimental system than *X. laevis*, but we thought it worth the effort, as it provided the most direct approach to prove, or disprove, the hypothesis. In addition, mouse IVF offered several advantages to achieve better control of experimental conditions, including, among others, the possibility to use purified epididymal spermatozoa instead of sperm derived from chopped *X. laevis* testis, as well as a fertilization medium of well-controlled composition instead of tap water.

In a very simple experimental layout, mouse epididymal spermatozoa were first incubated with pSV_2CAT plasmid DNA (the plasmid drives the expression of the bacterial chloramphenicol acetyl-transferase (CAT) reporter gene under the control of the SV40 promoter); after incubation, the spermatozoa were used to fertilize oocytes in IVF assays. Fertilized oocytes were cultured *in vitro* to the two-cell stage and the next day embryos were implanted into foster mothers. After birth, DNA samples extracted from the tail of the animals were analysed by Southern blotting using CAT sequences as probes. After about one year of experiments, the results were reported in an article published in *Cell* [16], highlighting essentially four main points:

i.   mouse mature spermatozoa are indeed spontaneously able to bind exogenous DNA molecules;

ii. exogenous DNA sequences are delivered from the DNA-incubated sperm to oocytes at fertilization, are propagated throughout embryogenesis and are eventually inherited in tissues of F0 progeny;

iii. the exogenous CAT gene is expressed in some, though not in all, tissues of F0 individuals;

iv. the exogenous sequences can be sexually transmitted from F0 founders to F1 progeny.

Together, these results fulfilled the prediction that spermatozoa can be vectors of foreign genetic information and showed for the first time the potential of a sperm-mediated process to introduce new genetic and phenotypic traits in mammals. This process was called Sperm Mediated Gene Transfer (SMGT). Almost concomitant with our work, another article showed that sea urchin spermatozoa can also take up pSV2CAT plasmid DNA and deliver it to oocytes at fertilization, generating CAT-expressing swimming blastulae [17]. Both publications converged on the conclusion that the ability of spermatozoa to bind and transfer foreign DNA is a well-conserved feature in two evolutionarily distant species, respectively from echinoids and mammals.

Soon after the publication of our work, we realised that we had not been the first in the field: a pioneering article [18] was brought to our attention, which had previously shown that rabbit spermatozoa, depleted of seminal fluid, could take up SV40 DNA and pass it to oocytes at fertilization. In that study, the presence of the viral DNA was not revealed as generating a novel trait in the offspring, but it was detected in one- and two-cell stage-embryos by the cytotoxic effect which the embryos exerted on co-cultured CV1 cells.

In the pre-internet era, the work by Bracket and colleagues, of which we were unaware at the time we had published our own study, had remained surprisingly ignored by the scientific community and had not even been pursued by its authors, despite of its far-reaching implications. At that point, therefore, the two 1989 articles and the early Bracket study provided independent and consistent support to the notion that SMGT is a real experimentally testable phenomenon, with far-reaching applications in biotechnology and, if proved to occur in nature, with evolutionary implications.

THE CONTROVERSY ON SMGT

The publication of our *Cell* paper in 1989 had an impact on the scientific community, which did respond promptly with numerous attempts to reproduce the SMGT protocol. To our amazement, however, a great deal of controversy burst up when several laboratories reported their failure to generate transgenic mice using SMGT. The negative results were essentially derived from Southern blot analysis of DNA samples extracted from tail fragments of founder mice; the data were collected by Ralph Brinster and Richard Palmiter and reported in a letter to *Cell* [19], in which the authors pointed out that there was no simple solution for making transgenic mice using the SMGT protocol. The sudden burst of the controversy was totally unexpected to us. In those days we were not aware of any condition that could have cast doubt on our published data. Therefore, in our reply to Brinster and Palmiter's letter [20], we could not but confirm our original findings, yet we also had to face the evidence that a problem of reproducibility did evidently exist, which we had been completely unaware of and of no simple rationalisation.

The next several months passed in the effort to metabolize the emotional shock caused by the suddenly exploded controversy. We resumed our IVF trials in an effort to rule out the possibility that trivial experimental artifacts caused the lack of reproducibility (*e.g.*, the possible interference of phenol red in the medium with the DNA binding to sperm cells, the effects of the incubator CO_2 concentration on the sperm/DNA mix, *etc.*). After reassuring ourselves that this was not the case, we took the decision to undertake a systematic study of the mechanism underlying SMGT. That decision in 1990 opened a new phase in the story. The *Cell* paper had reported the empirical observation that spermatozoa can take up exogenous DNA and deliver it to oocytes at fertilization, but nothing was actually known on the mechanisms, molecular factors or biochemical parameters underlying these events. It became clear by that point that clarifying this mechanism, and identifying the key factors and parameters involved, was indispensable to rationalize the protocol, acquire full experimental control over the process and find rational solutions to the question raised by the scientific community. Though not strictly related to the clarification of the mechanism, additional puzzling questions were: how does this happen? Why spermatozoa that carry the paternal genome are spontaneously permeable to foreign genetic information, when this uptake can seriously compromise the genetic identity of the progeny? Are spermatozoa protected from undesired foreign intrusions and, if so, how? In order to answer these questions, we undertook a systematic dissection of the process, starting from its first and, in my view, most fundamental event, namely the interaction between spermatozoa and exogenous DNA.

THE END OF A PREJUDICE: LIVING SPERMATOZOA ARE HIGHLY ACCESSIBLE TO FOREIGN DNA

The decision to systematically tackle the mechanism underlying the ability of sperm cells to bind foreign molecules opened up a new phase that continued over the following ten years. In essence, the work done in that phase indicated that the binding between spermatozoa and exogenous DNA is not a casual event but is a well-regulated interaction, mediated by specific protein factors on the outer spermatozoa surface; the exogenous DNA that becomes sperm-bound can be further internalized into nuclei and becomes embedded within sperm chromatin, in close contact with chromosomal DNA. Detailed descriptions of these findings can be found in earlier reviews [21-23] and in other chapters of this book. Here, I will only briefly recall the following relevant aspects:

- foreign DNA molecules bind preferentially on the sub-acrosomal segment of the sperm head, regardless of the sperm cell shape and of the animal species used as sperm donors. The binding is ionic, reversible, sequence-independent, not restricted to DNA only but capable of occurring with many negatively charged macromolecule [24].

- The binding is mediated by a group of 30-35 kDa DNA-binding proteins, located on the sperm surface, acting as 'receptors' for exogenous nucleic acid molecules [25].

- The binding of DNA to sperm cells is inhibited in the presence of seminal fluid, which contains an inhibitory factor (IF-1) with high affinity for the DNA receptor proteins [24, 25]. The binding of IF-1 to the 30-35 kDA proteins prevents the interaction of exogenous DNA molecules. Therefore, only epididymal or ejaculated and thoroughly washed spermatozoa fully depleted of seminal fluid can bind DNA, whereas spermatozoa in ejaculated semen are virtually inaccessible. Noteworthily, these results provided an answer to a crucial question, because they pinpointed a natural mechanism of protection exerted by the seminal fluid over ejaculated spermatozoa against intrusions of foreign DNA molecules when sperm cells are in the female genital tract (mammals) or spawn in the environment (aquatic animals). We later found that a second level of protection is provided by endogenous nucleases(s) that are activated when spermatozoa are exposed to exogenous DNA. The endogenous nuclease response is triggered in a DNA dose-dependent manner and causes the degradation of the sperm-bound foreign molecules and, in certain cases, the death of the spermatozoa cells themselves through a process that resembles apoptosis [26].

Moving to the next step of the process, we asked what happens to the DNA that becomes sperm-bound. We found that a fraction of this DNA is internalized in nuclei in close contact with the chromosomal DNA [27]; the nuclear internalization is a CD4-dependent process [28]. Following internalization, rare events of recombination can take place, that can cause the integration of foreign sequences in the sperm genomic DNA, albeit with very low frequency [29].

On the whole, these findings somehow subverted the traditional view of spermatozoa as tightly packed, metabolically inert cells and revealed that these cells are in fact endowed with a network of unsuspected functions mediated by specific factors. From an applied point of view, the discovery of this network provided a mechanistic foundation for SMGT and for our empirical observations; in a broader biological perspective, it provided consistent clues to suggest that spermatozoa are complex cells, expressing many more functions than previously believed and involved in many more activities than simply the delivery of the male genome.

EXPANSION OF THE SMGT FIELD

Parallel to our work, a large number of laboratories worldwide were making substantial contributions to the SMGT field and extended their studies to a variety of animal species: these articles, first appearing in the early '90s and progressively increasing in the following years, have unanimously provided sound confirmatory bases for the ability to bind exogenous DNA as a general feature of spermatozoa from virtually all species (Chapter 2 of this book; also see [22]). Only a small number of articles, however, have reported on studies with mouse spermatozoa, while the majority came from laboratories working with a variety of different species, especially acquatic animals. I believe that the controversy exploded after our 1989 publication has discouraged efforts in SMGT for most of those working in mouse transgenesis; this would be expected, also considering that microinjection-based transgenesis works with a pretty high efficiency in mice, while having a very poor outcome in other species, including those of commercial interest [30]. This probably justifies the willingness of those working with those species to invest efforts to try and develop new transgenesis technologies, including SMGT, than those working in mouse.

Since its first description, SMGT has been perceived as a potential biotechnological tool, providing an easy and low-cost method to create transgenic animals and avoiding the tedious and technically demanding DNA microinjection procedure. Consistent with this view, most publications in the '90s focussed on the applied side of SMGT aiming at the production of transgenic animals, whereas studies on its possible role(s) in nature and implications for evolution and human health when exogenous molecules are inadvertently taken up by sperm cells, remained comparatively underdeveloped - though with some remarkable exceptions [31-35]. The results from SMGT reports almost unanimously converged on three main conclusions:

i. spermatozoa of virtually all animal species bind spontaneously exogenous DNA molecules;

ii. foreign DNA is transferred from spermatozoa to oocytes and further propagated to developing embryos;

iii. the final fate of the foreign molecules is extremely heterogeneous.

A broad spectrum of outcomes has actually been described: foreign DNA sequences can occasionally integrate in the host genome or (more often) remain extrachromosomal; they can be propagated to adult animals or be only present in embryos; they can be mosaic or homogeneously distributed in the tissues, transcriptionally competent or silent, transmitted to the F1 progeny (in a non-Mendelian fashion, in this case) or detected only in F0 founders. The cause of such heterogeneity has long remained obscure, as no clear link to any identifiable experimental parameter could be established, nor was it distinctive of a specific animal system.

An innovative article from the group of Yanagimachi [36] provided a first breakthrough into this question in 1999. The authors showed that the combination of SMGT and ICSI (intracytoplasmic sperm injection), which entails microinjecting DNA-loaded demembranated spermatozoa in the oocytes, dramatically improved the production of genuine transgenic mice. That seminal work, the first of a series that followed, pointed out the key role of the plasma membrane of spermatozoa for the final fate of exogenous DNA molecules: when intact spermatozoa take up DNA and deliver it to oocytes, as in conventional SMGT protocols, extrachromosomal structures are mostly generated; when demembranated sperm cells are first incubated with DNA and then injected in oocytes, instead, high number of transgenic animals are obtained (see chapter 2 of this book). The results of Perry and coworkers provided a brilliant solution to improve transgenic animal production using sperm cells as vectors of exogenous DNA and suggested an alternative protocol combining SMGT and ICSI.

Besides of its practical outcome, the Yanagimachi work had crucial basic implications, as it raised a novel fundamental question: if transgenesis does require the direct contact of exogenous DNA with nuclear chromatin, what then are the real nature and origin of the new traits seen in offspring when DNA is taken up by intact spermatozoa? An analysis of the animals obtained with the two approaches suggests that new traits acquired through SMGT protocols are extremely heterogeneous and unpredictable (*e.g.* in intensity, distribution, frequency and persistence within the same animal). In contrast, genuine transgenes are stably integrated in the host genome and transmissible from one generation to the next as Mendelian characters. In other words, the work carried out in this phase led to the conclusion that SMGT and transgenesis reflect two distinct biological processes with different outcomes.

Building up on that notion, a vast body of currently available data fully confirms that SMGT can indeed induce the genesis and propagation of new genetic and phenotypic traits in animals, on the one hand; on the other hand, it does not entail stable integration in the genome and therefore cannot provide clear predictions on the true genetic identity or fate of these animals in embryonic development or in adulthood, given that the genetic outcome remains unpredictable and heterogeneous. To date, therefore, fundamental aspects of the mechanism remain unknown, which need further studies to characterize the real nature of the newly generated traits and also to gain full experimental control on the process.

SPERM AND RETROELEMENTS, A PRODUCTIVE RELATION: THE RETROGENE FACTORY

Another decisive turning point of the story took place in 2000, with the finding in my laboratory that mouse mature spermatozoa are endowed with a reverse transcriptase (RT) activity [37]. RT activity was identified on the basis of the ability of sperm cells to reverse-transcribe exogenous RNA molecules in cDNA copies. Precursor of this finding was the characterisation of a minor component of mouse sperm chromatin that was found to be highly enriched in retrotransposon DNA [38], and was organised in nucleosomes - rather than in protamine complexes - nuclease-sensitive, hypomethylated [39] and associated with transcription factors [40]. Those features suggested that this chromatin fraction is organised in an "active" conformation in mature spermatozoa and that retrotransposon sequences, abundant in this fraction, may have been expressed during the spermatozoa maturation process. That was the background that had led us to investigate and identify RT, an enzyme encoded by retroelements, in sperm cells [37]. The finding of a functional RT in sperm cells was unexpected and raised puzzling questions on its role, if any. This question grew more compelling when we found that RT is strictly required for early embryo developmental stages, as RT inhibition (by either pharmacological inhibitors [41] or down-regulating the RT-encoding LINE-1 retrotransposon family by LINE-1 antisense oligonucleotide microinjection in male pronuclei of mouse zygotes [42]), irreversibly arrest embryonic development at the two- and four-cell stages.

An unexpected answer to the question of the role of RT in sperm cells came some years later. To our own surprise, we found that when foreign RNA was used instead of DNA in SMGT and taken up by spermatozoa, biologically active "retrogenes" are spontaneously generated *via* reverse transcription [43]. More specifically, we were able to show that epididymal spermatozoa incubated with an RNA population, transcribed from a beta-gal-containing construct, generate cDNA copies that are delivered to oocytes, propagated in the developing embryos and further in tissues of born individuals, where the reporter gene is eventually mosaic expressed in some but not all tissues [43]. We further demonstrated that this same process also occurs when spermatozoa are preincubated with a tagged DNA construct: in this case, the DNA triggers a two-steps reaction in which reverse transcription follows a prior transcription step that generates RNA from the original DNA construct [44]. The resulting reverse-transcribed sequences essentially remain as low-copy, non-integrated episomal structures, that can be inherited in a non-mendelian fashion and even expressed in some tissues of F1 individuals (more details are given in Chapter 12 of this book; also see [45]). Therefore, regardless of whether sperm cells are incubated with exogenous RNA or DNA molecules, transcriptionally competent retrogenes can be generated, propagated during development and in adult tissues, and transmitted from one generation to the next, as a continuous source of novel phenotypic traits even in the absence of corresponding chromosomal genes.

CONCLUSIONS

To sum up, I have presented a personal recapitulation of SMGT studies, spanning over a period of twenty years, which I ideally divide in three distinct phases: i) a first phase, starting in the late '70s, with the early observations of sperm permeability to nucleases, and concluded in 1989 with our first article reporting on SMGT; ii) a second phase, started in 1990 and lasted almost ten years, characterized by progress in clarification of relevant aspects of the mechanism of DNA/binding and internalization and culminating with the finding that the combination SMGT/ICSI improves transgenesis; iii) a third phase, characterized by the discovery that spermatozoa are endowed with an endogenous RT activity which is a key component of an active mechanism responsible for the genesis of new genetic information. The first two phases and the body of generated results have been consistent with their original inspiring questions on the possible use of sperm cells as vectors of foreign information. In the last phase, instead, the work has taken an entirely new turn, with hints to novel directions apparently unrelated to the original premises.

As recalled above, our goal in undertaking a systematic dissection of the SMGT process twenty years ago was to confirm the consistency of our data, to clarify the reasons for the apparent lack of reproducibility of the protocol and to provide an experimental foundation to the empirical observation that spermatozoa are vectors of exogenous genetic information. To our surprise, the newly acquired experimental data in the last twenty years, while fully confirming the suitability of sperm cells, have also disclosed a novel scenario, well beyond our initial views and expectations. SMGT now emerges not only as a mere carriage of foreign DNA by spermatozoa, as we initially thought, but as a more complex phenomenon implying the existence of a previously unknown RT-mediated mechanism which confers to spermatozoa the power to be a continuous source of novel genetic and phenotypic extrachromosomally inherited traits. These traits provide a subtle flow of genetic information, independent from that encoded by chromosomal genes: some of these features share close analogies with the recently reported epigenetic phenomenon of RNA-mediated inheritance [46, 47]. Whether this phenomenon occurs in nature, potentially affecting the health, the genetic identity and the fate of species including human, will be a matter of future investigation.

REFERENCES

[1]   Avery OT, Macleod CM, McCarty M. Studies on the chemical nature of the substance inducing transformation of pneumococcal types. J Exp Med 1944; 79: 345-59.

[2]   Watson JD, Crick FH. Molecular structure of nucleic acids; a structure for deoxyribose nucleic acid. Nature 1953; 171: 737-38.

[3]   Keeling PJ, Palmer JD. Horizontal gene transfer in eukaryotic evolution. Nat. Rev 2008; 9: 605-18.

[4]   Gladyshev EA. Meselson M., Arkhipova IR. Massive Horizontal Gene Transfer in Bdelloid Rotifers. Science 2008; 320: 1210-13.

[5]   Frees KK, Tuveson DA. Maximizing mouse cancer models. Nat Rev Cancer 2007; 7: 645-58.

[6]   Haruyama N, Cho A, Kulkami AB. Overview: engineering transgenic constructs and mice. Curr Protoc Cell Biol 2009; Chapter 19: Unit 19.10.

[7]   Houdebine LM. Transgenic animals bioreactors. Transgenic Res 2000; 9: 305-20.

[8]   Redwan el-RM. Animal-derived pharmaceutical proteins. J Immunoassay Immunochem 2009; 30: 262-90.

[9]   Gross DS, Garrard WT. Nuclease hypersensitive sites in chromatin. Ann Rev Biochem 1988; 57: 159-97.

[10]   Telford DJ, Stewart BW. Micrococcal nuclease: its specificity and use for chromatin analysis. Int J Biochem 1989; 21: 127-37.

[11]   Kornberg RD. Structure of chromatin. Ann Rev Biochem 1977; 46: 931-54.

[12]   Spadafora C, Geraci G. The subunit structure of sea urchin sperm chromatin: a kinetic approach. FEBS Lett 1975; 57: 79-82.

[13]   Spadafora C, Compton L, Gross-Bellard M, *et al.* The DNA repear lengths in chromatins from sea urchin sperm and gastrulae cells are markedly different FEBS Lett 1976; 69: 281-85.

[14]   Geraci G, Noviello L. DNase hydrolysis of chromatin DNA in intact sea urchin sperm cells. Effects of the ionic strength on the digestion parameters. Cell Diff 1979; 8: 203-10.

[15]   Whittingham DG. Culture of mouse ova. J Reprod Fert Suppl 1971; 14: 7-21.

[16]   Lavitrano M, Camaioni A, Fazio V, *et al.* Sperm cells as vectors for introducing foreign DNA into eggs: genetic transformation of mice. Cell 1989; 57: 717-23.

[17]   Arezzo F. Sea urchin sperm as a vector of foreign genetic information. Cell Biol Int Rep 1989; 13: 391-04.

[18]   Bracket BG, Boranska W, Swicki W, *et al.* Uptake of heterologous genome by mammalian spermatozoa and its transfer to ova through fertilization. Proc Natl Acad Sci USA 1971; 68: 353-57.

[19]   Brinster RN, Sandgren EP, Behringer RR, *et al.* No simple solution for making transgenic mice. Cell 1989; 59: 239-41.

[20]   Lavitrano M, Camaioni A, Fazio V, *et al.* No simple solution for making transgenic mice: reply Cell 1989; 59: 241.

[21] Spadafora C. Sperm cells and foreign DNA: a controversial relation. BioEssays 1998; 20: 955-64.

[22] Smith KR, Spadafora C. Sperm-mediated gene transfer: applications and implications. Bioessays 2005; 27: 551-62.

[23] Niu Y, Liang S. Progress in gene transfer by germ cells in mammals. J Genet Genomics 2008; 35: 701-14.

[24] Lavitrano M, French D, Zani M, *et al.* The interaction between exogenous DNA and sperm cells. Mol Reprod Dev 1992; 31: 161-69.

[25] Zani M, Lavitrano M, French D, *et al.* The mechanism of binding of exogenous DNA to sperm cells: factors controlling the DNA uptake. Exp Cell Res 1995; 217: 57-64.

[26] Maione B, Pittoggi C, Achene, *et al.* Activation of endogenous nucleases in mature sperm cells upon interaction with exogenous DNA. DNA & Cell Biol 1997; 16: 1087-97.

[27] Francolini M, Lavitrano M, Lora-Lamia C, *et al.* Evidence for nuclear internalization of exogenous DNA into mammalian sperm cells. Mol Reprod Dev 1993; 34: 133-39.

[28] Lavitrano M, Maione B, Forte E, *et al.* The interaction of sperm cells with exogenous DNA: a role of CD4 and major histocompatibility complex class II molecules. Exp Cell Res 1997; 233: 56-62.

[29] Zoraqi G, Spadafora C. Integration of foreign DNA sequences into mouse sperm genome. DNA & Cell Biol 1997; 16: 291-00.

[30] Robl JM, Wang Z, Kasinathan P, *et al.* Transgenic animals production and animal biotechnology. Theriogenology 2007; 67: 127-33.

[31] Baccetti B, Benedetto A, Burrini AG, *et al.* HIV-particles in spermatozoa of patients with AIDS and their transfer to oocytes. J Cell Biol 1994; 127: 903-14.

[32] Bobroski L, Bagasra AU, Patel D, *et al.* Localization of human herpes virus type 8 [HHV-8] in the Kaposi's sarcoma tissues and the semen specimen of HIV-1 infected and uninfected individuals by utilizing *in situ* polymerase chain reaction. J Reprod Immunol 1998; 41: 149-60.

[33] Smith KR. The role of sperm-mediated gene transfer in genome mutation and evolution. Med Hyp 2002; 59: 433-37.

[34] Nuno Moreira P, Fernandez-Gonzales R, Rizos D, *et al.* Inadvertent transgenesis by conventional ICSI in mice. Hum Reprod 2005; 20: 3313-17.

[35] Ceballos A, Remes Lenicov F, Sabatte J, *et al.* Spermatozoa capture HIV-1 through heparan sulfate and efficiently transmit the virus to dendritic cells. J Exp Med Oct 2009; 26.

[36] Perry AC, Wakayama T, Kishikawa H, *et al.* Mammalian transgenesis by intracytoplasmic sperm injection. Science 1999; 284: 1180-83.

[37] Giordano R, Magnano AR, Zaccagnini G, *et al.* Reverse transcriptase activity in mature spermatozoa of mouse. J Cell Biol 2000; 148: 1107-13.

[38] Pittoggi C, Renzi L, Zaccagnini G, *et al.* A fraction of mouse sperm chromatin is organized in nucleosomal hypersensitive domains enriched in retroposon DNA. J Cell Sci 1999; 112: 3537-48.

[39] Pittoggi C, Zaccagnini G, Giordano R, *et al.* Nucleosomal domains of mouse spermatozoa chromatin as potential sites for retroposition and foreign DNA integration Mol Reprod Dev 2000; 56: 248-51.

[40] Pittoggi C, Magnano AR, Sciamanna I, *et al.* Specific localization of transcription factors in the chromatin of mouse mature spermatozoa. Mol Reprod Dev 2001; 60: 97-06.

[41] Pittoggi C, Sciamanna I, Mattei E, *et al.* A role of endogenous reverse transcriptase in murine early embryo development. Mol Reprod Dev 2003; 66: 225-36.

[42] Beraldi R, Pittoggi C, Sciamanna I, *et al.* Expression of LINE-1 retrotrasposons is essential for murine preimplantation development. Mol Reprod Dev 2006; 73: 279-87.

[43] Sciamanna I, Barberi L, Martire A, *et al.* Sperm Endogenous Reverse Transcriptase as Mediator of New Genetic Information. Biochem Bioph Res Comm 2003; 312: 1039-46.

[44] Sciamanna I, Vitullo P, Curatolo A, *et al.* Retrotransposons, reverse transcriptase and the genesis of new genetic information. Gene 2009; 448: 180-6.

[45] Pittoggi C, Beraldi R, Sciamanna I, *et al.* Generation of biologically active retro-genes upon interaction of mouse spermatozoa with exogenous DNA. Mol Reprod Dev 2006; 73: 1239-46.

[46] Spadafora C. Sperm-mediated 'reverse' gene transfer: a role of reverse transcriptase in the generation of new genetic information. Hum Reprod 2008; 23: 735-40.

[47] Rassoulzadegan M, Grandjean V, Gounon P, *et al.* RNA-mediated non-mendelian inheritance of an epigenetic change in the mouse. Nature 2006; 441: 469-74.

CHAPTER 2

SMGT Research Findings: An Overview

Ilaria Sciamanna and Corrado Spadafora[*]

Italian National Health Institute, Viale Regina Elena 299, 00161 Rome, Italy

Abstract: The finding that spermatozoa of virtually all animal species can spontaneously bind exogenous DNA molecules and deliver it to oocytes at fertilisation first suggested that these cells can be used as vectors for introducing new genetic and phenotypic traits in animals. That has led to the development of a novel approach to animal transgenesis, namely Sperm Mediated Gene Transfer (SMGT). Here we review findings obtained using this experimental approach. A critical examination of published evidence indicates that the alternative between direct binding to the plasma membrane of sperm cells, or its bypass, represents a crucial parameter for the fate of exogenous nucleic acid molecules: in the former case, episomal structures are mainly generated; in the latter, integration in the host genome is more frequent. The original protocol was based on the direct interaction between sperm cells and foreign DNA. Several alternative variants have been developed thereafter to improve the efficacy of the method. Improved protocols include the combination of SMGT with: i) ICSI (intracytoplasmic sperm injection) technology, ii) restriction enzymes favoring DNA integration (REMI), or iii) linker-based (LB), in which the DNA binding is mediated by an antibody recognizing a membrane antigen. In another approach, the aim is to produce "transgenic spermatozoa": for example, in testis-mediated gene transfer (TMGT) the foreign DNA is microinjected directly into testis; in virus-mediated transgenesis, new genes are delivered to spermatogonial stem cells by viral vectors. The recent finding that mature spermatozoa are the source of non-integrated, transcriptionally competent retrogenes also suggests a potential use of SMGT for embryonic gene therapy.

Keywords: SMGT, Transgenesis, Episomal structures, DNA integration, RNA-mediated transgenesis, ICSI, REMI, TMGT, Non-Mendelian characters, Sperm-mediated gene therapy.

INTRODUCTION

Spermatozoa are traditionally viewed as metabolically inert cells, because they lack most fundamental biochemical and molecular functions typical of somatic cells. In sharp contrast with this view, new findings obtained in recent years have subverted this view and have revealed that sperm cells are capable of a variety of previously unsuspected metabolic functions.

In spite of their apparently inert transcriptional state, spermatozoa in fact store a large repertoire of heterogeneous RNA populations (reviewed in [1-3]) characterized by a complex profile, estimated to include several thousands of mRNAs [4] and miRNAs [3, 5, 6]. The identification of transcribed sequences in mature spermatozoa is consistent with the finding that a collection of transcription factors is also present in these cells [7]. Interestingly, a proportion of this RNA is thought to be translated within sperm cells through the mitochondrial ribosomal apparatus [8]. Another part is delivered to oocytes at fertilisation [9], implying that spermatozoa contribution to zygotes is not only restricted to the paternal genome, but also includes RNA-based information. Indeed, it is now accepted that spermatozoa play a more complex role than previously thought in the passage of genetic information at fertilisation. Such a complex role is elicited, for example, by the finding that spermatozoal RNA molecules can mediate non-mendelian inheritance of epigenetic traits in mice [6, 10]. Furthermore, a sperm endogenous reverse transcriptase-dependent machinery can generate and propagate transcriptionally competent retrogenes throughout development [reviewed in 11, 12].

Sperm cells can be used as delivery vectors of foreign genetic information in transgenesis and their spontaneous ability to bind exogenous DNA or RNA molecules constitutes the foundation for SMGT (see

*Address correspondence to Corrado Spadafora: Italian National Health Institute, Viale Regina Elena 299, 00161 Rome, Italy; Tel: 39 (0)6 49903117; E-mail: corrado.spadafora@iss.it

chapters 1 and 12 of this book; also reviewed in [13-15]). The original protocol developed in our laboratory in the mouse system employed sperm cells that interacted directly with exogenous DNA molecules [16]. Variant SMGT protocols have subsequently been developed in numerous animal species, with the dual aim to adapt the protocol to the specific reproductive requirements of different species and to improve the efficiency of the method.

The attempts to define and establish reliable and effective SMGT protocols applicable to different animal species are of high value in biotechnology. DNA microinjection requires costly and sophisticated equipment and a great manual skill in micromanipulation. In addition, the microinjection technology, though successfully applied to mice, works far less efficiently with other species, such as livestock and marine animals. SMGT can represent an advantageous alternative for animal species that are refractory to microinjection. In addition, the SMGT procedure does not need any particular equipment or skill and can be easily performed in field work. Another attractive aspect of using spermatozoa as DNA vectors lies in the possibility of carrying out mass transgenesis. In contrast with microinjection, which needs performing in each individual embryo, genetic transformation of a large number of embryos can be collectively achieved in one step by SMGT. This can be of particular interest in the transformation of marine species. On those grounds, SMGT is perceived in a somewhat restricted biotechnological perspective, essentially as a simplified, easy-to-handle and affordable method for the production of trangenic animals compared to DNA microinjection into the zygotic pronucleus. Comparatively less attention has been paid to basic aspects, *e.g.* the implication of sperm cells as vectors of heterologous genetic information, the molecular mechanism mediating the interaction as well as the internalisation of foreign DNA or RNA, and the ability of spermatozoa to provide a source of novel genetic information in nature. Studies addressing these issues have disclosed novel functions that have expanded the implications of these cells – well beyond the transgenesis field – in aspects of genetics and non-Mendelian transgenerational inheritance (see Chapters 1 and 12 of this book; also references [12-15]). The purpose of the present chapter is to provide an updated overview of the results from both basic studies of the mechanism controlling sperm/DNA interaction and from the SMGT field aiming at the production of animal strains expressing novel genetic traits.

MECHANISM OF SPERM/DNA INTERACTION AND NUCLEAR INTERNALISATION

It is now well established that spermatozoa of virtually all animal species can take up exogenous nucleic acid molecules and internalize them into nuclei, as summarized in Table **1**.

Table 1: Studies Reporting the Interaction of Exogenous DNA with Spermatozoa of Various Species

Class	Species	References
Echinoids	Sea Urchin	Arezzo F. 1989 [18]
Molluscs	Abalone	Tsai *et al.* 1997; Chen *et al.* 2006 [53]
Insects	Blow fly, Honey bee, Silkworm, Lucilla cuprina, Apis melifera	Atkinson *et al.* 1991; Shamila and Mathavan, 2000; Robinson *et al.* 2000;
Fish	Carp, African catfish, Tilapia, Zebrafish, Salmon, Loach, Sea bream	Zhong *et al.* 2002; Muller *et al.* 1992; Kurita *et al.* 2004; Khoo, 2000; Patil *et al.* 1996; Sin *et al.* 2000; Sin *et al.* 1993; Chen *et al.* 1997; Lu *et al*, 2002
Amphibians	Xenopus laevis	Jonak *et al.* 2000; Sparrow *et al.* 2000; Smith *et al.* 2006; L'hostis-Guidet *et al.* 2009;
Birds	Rooster	Nakanishi and Iritani, 1993; Harel-Markowitz *et al.* 2009

Table 1: cont….

	Mouse,	Lavitrano *et al.* 1989; Sciamanna *et al.* 2000;
	Rat,	Hirabayashi *et al.* 2010; Yonezawa *et al.* 2002; Hibbitt *et al.* 2006;
	Hamster,	Brackett *et al.* 1971; Li *et al.* 2006; Li *et al.* 2005; Kurome *et al.* 2006;
	Rabbit,	Hoelker *et al.* 2007;
	Goat,	Moisayadi *et al.* 2009;
Mammals	Pig,	Sciamanna *et al.* 2000;
	Bull,	Shemesh *et al.* 2000;
	Horse,	Ball *et al.* 2008;
	Rhesus monkey	Chan *et al.* 2000;
	Human	Camaioni *et al.* 1992

Spermatozoa typically bind exogenous DNA in the subacrosomal segment of the sperm head (as shown in Fig. **1** for murine spermatozoa), regardless of the morphology of sperm cells - whether rounded or hook-shaped. The molecular mechanisms underlying the interaction between sperm cells and foreign DNA and its subsequent internalisation are described in depth elsewhere [11, 13-15].

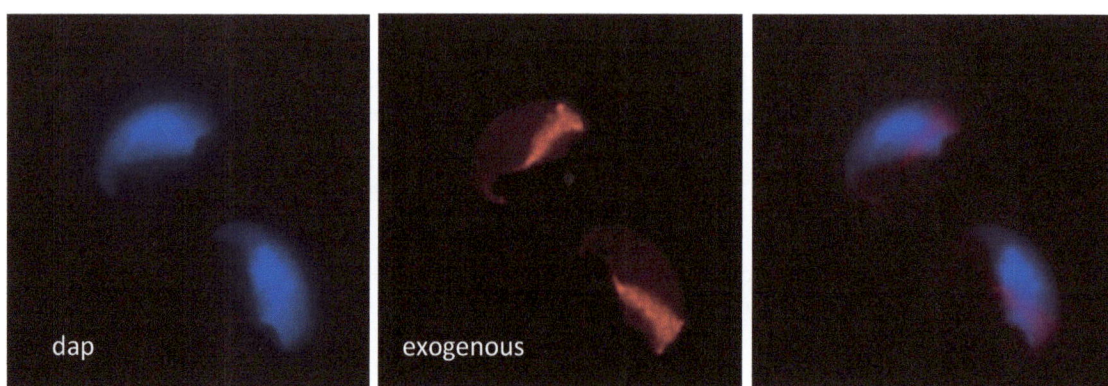

Figure 1: The Interaction of Exogenous Plasmid DNA with Mouse Epididymal Spermatozoa. Left panel: DAPI staining of sperm DNA. Middle panel: sub-acrosomal localization of biotin- labelled foreign DNA revealed with Cy3-conjugated streptavidin. Right panel: merge of the DAPI and biotin images.

A network of activities are implicated in the processes summarized in Box 1, indicating that the interaction between DNA and spermatozoa is a complex and well-regulated process, whose scope in nature is not fully understood. The binding of exogenous DNA and its internalisation in the nuclei of sperm cells are key steps in the process of SMGT. Modified protocols have been developed to improve these steps, including, among others, electroporation and lipofection of sperm cells. Electroporation has been successfully used with spermatozoa of a variety of vertebrate and invertebrate species to facilitate the uptake of exogenous DNA molecules [13, 14].

Box 1. Factors and Processes in the Internalisation of Exogenous DNA in Spermatozoa: a Summary.

- Exogenous DNA can interact with either epididymal, or ejaculated and thoroughly washed spermatozoa. The presence of seminal fluid strongly antagonizes the interaction, because it contains a glycoprotein (termed IF-1, Inhibitory Factor 1), abundant in the seminal fluid of mammals and on the surface of lower eukaryote spermatozoa, that exerts a powerful inhibitory effect on DNA interaction, thus protecting spermatozoa from undesired intrusions in nature.

- The binding of exogenous DNA to sperm cells is mediated by a group of 30-35 kDa proteins located on the sperm cell membrane, acting as DNA-binding substrates (DBS); the assembly of DBS/DNA complexes is strongly inhibited by the IF-1 glycoprotein.

> - The interaction with DBSs triggers a process of nuclear internalisation, culminating with the internalisation of a constant proportion (15-22%) of sperm-bound DNA in nuclei [67]. That observation suggests that the internalisation is not the consequence of passive transfer, but is mediated by a specific intake mechanism.
>
> - The transmembrane glycoprotein CD4 present on spermatozoa plays a key role in internalisation: indeed, spermatozoa from CD4 knock-out mice, though being able to bind DNA, fail to internalise it into nuclei [65].
>
> - After the internalisation, the DNA/DBS/CD4 complex dissociates and the exogenous sequences are released in close contact with the nuclear scaffold.
>
> - Therein, the exogenous sequences eventually undergo recombination and, in rare cases, can integrate into the sperm genome [13, 22]. The internalisation of exogenous DNA triggers the activation of sperm endogenous nuclease activities in a dose-dependent manner.

These nucleases degrade the exogenous DNA [66], a phenomenon that may be viewed as a 'second line' defense - after that opposed by the inhibitory factor IF-1 - in response to the danger of massive invasion of spermatozoa by foreign DNA.

Increased rates of exogenous DNA internalisation have also been obtained by preincubating spermatozoa with exogenous DNA constructs in the presence of cationic lipids (liposomes), with no apparent toxicity or alteration of the functional activity of sperm cells. A simple transfection method that uses magnetic nanoparticles (MNPs) has recently been reported [17]. The complexes formed by plasmid DNA and MNPs showed a higher efficiency of association with ejaculated boar spermatozoa compared to naked DNA or DNA:liposome mixtures. What follows is an overview of the experimental strategies developed to transfer DNA molecules into oocytes at fertilisation by exploiting the ability of spermatozoa to interact with exogenous DNA.

SPERM-MEDIATED GENE TRANSFER AFTER DIRECT INTERACTION OF SPERM CELLS WITH EXOGENOUS DNA

In the original SMGT procedure, the exogenous DNA was allowed to interact directly and spontaneously with mature spermatozoa, and the sperm/DNA complex was then used for *in vitro* fertilisation (IVF) or artificial insemination (AI) assays. Because the ability to bind and internalize exogenous DNA is a widespread feature of spermatozoa regardless of their origin, this procedure has been extended to a variety of animal species. Reproductive features however are highly divergent among species, *e.g.* compare marine animals and mammals. It would therefore be impossible to develop a universal SMGT protocol applicable to all species. There has in fact been a need to develop species-specific protocols matching the reproductive and physiological requirements of different species. Three main experimental strategies have been adopted:

i.   In the murine system, epididymal spermatozoa suspended in medium are incubated with plasmid DNA and then directly used in IVF to generate embryos. Healthy two-cell embryos are selected and surgically implanted into foster mothers.

ii.   In farm animals, *i.e.* cattle and swine, SMGT has adapted the procedures that breeders currently use for AI: the semen is freshly collected from donor animals and thoroughly washed in appropriate medium so as to deplete spermatozoa from the seminal fluid through sequential rounds of low-speed centrifugation. The suspension of thoroughly washed sperm cells is then incubated with exogenous plasmid DNA (1 hour at 18°C) diluted in the appropriate medium, and used for AI.

iii. Finally, in aquatic animals, *e.g.* fish and amphibia, SMGT consists of a simplified procedure: spermatozoa are surgically obtained from male donors, incubated with the exogenous DNA and then used to fertilise eggs by external insemination in tap or sea water. It has been shown that exposure of sea urchin sperms to a short hypotonic shock (0.8x diluted sea water) prior the incubation with the exogenous DNA improves the DNA uptake [18]. Presumably, the hypotonic step removes the IF-1 inhibitory factor (see Box 1) present in the sperm plasma membrane, hence facilitating the binding and internalisation of DNA molecules.

Table **2** summarizes the results of SMGT experiments based on direct sperm/DNA interaction in different species. The table essentially focusses on the features and fate of foreign sequences in adult animals and their progeny, because their molecular characterization is generally sound, whereas their presence in embryos is usually assessed by either PCR or expression methods, neither of which provides conclusive evidence for transgene integration. A highly heterogeneous scenario emerges from the results, with a broad range of different outcomes. Generally, the direct binding between sperm cells and foreign DNA yields new traits which, in many cases, are transmitted transgenerationally from founders to their progeny. Most often, however, the new genetic traits do not correlate with the presence of integrated transgenes and the phenotypic features of expressing individuals share little in common with those of genuinely transgenic animals. In-depth studies in the mouse system indicate that the outcomes generated *via* direct interaction between sperm and DNA exhibit very peculiar features:

  i.   They are generally transient, being abundant in embryonic development – most preimplantation embryos analyzed by PCR contain the exogenous sequences - and in young adults, but tend to disappear during ageing;

  ii. They are mosaically distributed in tissues of adult animals;

  iii. They are usually present in low copy number (below one copy/genome and therefore detectable by PCR but not readily by Southern blot analysis);

  iv. They are inherited through generations in a non-Mendelian fashion and their expression is extremely variable [19-21] (reviewed in [12]).

On the whole, the data support the conclusion that the direct DNA interaction protocol does not generate stable transgenic strains; indeed, the integration of exogenous sequences in the sperm genome is a rare event [22].

A few discordant examples from this general trend are present in the literature. In one case it has been claimed that transgenic pigs expressing hDAF are efficiently generated by SMGT after direct sperm/DNA interaction [23]. No compelling evidence was however provided to indicate whether the human transgene was integrated in the swine genome, nor are any data available to document whether the foreign sequences were stably maintained in both founders and F1 individuals. Futhermore, the authors's claim that their high rate of transgenesis is based on an accurate selection of sperm donor boars [24] is not supported by any established demonstration that a link exists between the semen quality and the rate of foreign sequence integration in the sperm genome.

In synthesis, a wealth of data converge to suggest that the direct interaction of sperm cells with the DNA can yield phenotypically transformed, yet not genuinely transgenic animals. On the other hand, SMGT studies have unexpectedly disclosed a novel area of non-Mendelian transgenerational genetics. This is so because the system appears to generate new phenotypic traits, unlinked to chromosomal genes, that are expressed in animal tissues while, paradoxically, the corresponding genes appear to be absent from the host genome (see Chapters 2 and 10; reference [12]).

Many novel experimental strategies have been explored in the attempt to overcome the heterogeneity of the outcomes obtained using direct DNA interaction protocols, some of which have made real breakthroughs in developing alternative SMGT protocols.

Table 2: SMGT by Direct Incubation: Fate of the Exogenous DNA after IVF

Species	Stage of Detection of the Transgenes	Remarks	References
Rabbit	Embryos	Southern blot: DNA integration with major rearrangements	Brackett *et al.* 1971
Sea urchin	Embryos	No evidence for gene expression, Mendelian inheritance or genomic integration. Transmission of foreign DNA through generations was observed	Arezzo, 1989
Mouse	Viable animals		Lavitrano *et al.* 1989
Pig	Embryos	Foreign DNA mosaic distribution. Infrequent integration of foreign DNA and poor level of expression of the foreign gene	Horan *et al.* 1992
Zebrafish	Viable animals		Khoo *et al.* 1992
Salmon	Viable animals	Southern analysis reveals weak transgene-specific signals of aberrant size, below one copy/cell	Khoo, 2000
Mouse	Embryos		Sin *et al.* 1993, 2000
Bovine	Viable animals	Germline and somatic mosaicism. Integrated foreign DNA was modified by deletion and sometimes gradually eliminated in the course of ontogenesis. Very few transgenic adults obtained	Chan *et al.* 1995
Various mammals	Embryos		Schellander *et al.* 1995
Xenopus	Embryos, few adults	Analysis of somatic DNA identified 6% positive swine with major alterations of transgene DNA. Foreign sequences were revealed in sperm DNA of positive animals by PCR but not Southern blot	Chan *et al.* 1996
Bovine and pig	Embryos, adult animals		Habrova *et al.* 1996; Jonak, 2000
Mouse	Fetuses	Polarized results from 75 SMGT experiments: 13 variably positive (some with 100% efficiency), all other negative. Major DNA rearrangements in all positive samples.	Sperandio *et al.* 1996
Honey bee	Viable animals		Maione *et al.* 1998
Mouse, pig	Viable animals	No DNA integration by Southern blot. Non Mendelian transmission through generations	Robinson *et al.* 2000
Pig	Viable animals	Foreign sequences remain unintegrated and show heavy mosaicism, variable copy number and major rearrangements	Sciamanna *et al.* 2000
Rohu fish	Viable animals	Integration in 80% of pigs and transmission to progeny. correct transgene espression in 64%	Lavitrano *et al.* 2002; Webster *et al.* 2005
Mouse	Embryos, Fetuses,	Extrachromosomal low copy persistence, mosaic distribution, correct espression, non Mendelian inheritance from F0 to F1	Zhong *et al.* 2002
Pig	Adults		Pittoggi *et al.* 2006
Bovine	Fetuses	Foreign DNA remains extrachromosomal, low copy and mosaic distributed. Non-Mendelian inheritance from F0 to F1 with correct espression in both generations.	Manzini *et al.* 2006
Rabbit	Blastocysts		Alderson *et al.* 2006
Bovine	Embryos	A scaffold matrix attachment region-based vector remains episomal and is espressed in a high proportion of analyzed fetuses	Li *et al.* 2006
Pig	Embryos	PCR identification of foreign DNA in piglets at 3 days of age but not at 70-100 days of age. Detection of extrachromosomal transgenes in young offspring, but lost in adulthood	Hoelker *et al.* 2007
	Viable animals		Wu *et al.* 2008

IMPROVED SMGT PROTOCOLS

Successful SMGT protocols for animal transgenesis require three steps: DNA binding to the sperm vector, nuclear internalisation, and integration in the host genome (either the sperm genome, with the generation of "transgenic spermatozoa" or, after fertilisation, in the genome of one-cell embryos). Several strategies have been devised to improve these steps and obtain higher transgenic efficiencies: of these, ICSI (intracytoplasmic sperm injection), REMI (restriction enzyme-mediated integration) and LB-SMGT (linker-based SMGT) are the most common and are examined in more detail below. TMGT (testis-mediated gene transfer) and virus-mediated transgenesis adopt instead the different strategy of delivering transgenes to precursor spermatogonial cells either directly within the testis, or *in vitro* followed by implantation into the testis; both procedures activate the spontaneous production of "transgenic spermatozoa" and, subsequently, of transgenic animals *via* natural mating. Table **3** summarizes relevant results obtained using these variant protocols and, as Table **2**, focusses on the features and fate of exogenous sequences in born offspring and their progeny. Most of these alternative protocols unambiguously generate genuine transgenic animals and, in some cases, the rate of success is remarkably high.

Table 3: Summary of Combined Additional Methodologies and SMGT

Approach	Species	Outcome and Remarks	Reference
ICSI-Tr	Mouse adults Monkey embryos Pig embryos Swine adults Mouse adults Mouse adults Mouse adults	Transgenic embryos and adult F0 mice with correct expression of GFP reporter. Mendelian inheritance of the transgene from F0 to F1 progeny EGFP transgene expression in preimplantation embryos produced by ICSI, but not IVF Transgenic pigs expressing the reporter gene. Transgenic progeny from positive animal obtained by nuclear transfer Transgenic mice with stable incorporation (about 35%) and phenotypic expression of large yeast artificial chromosomes (YAC) constructs (250 Kbp). Mendelian inheritance of YAC transgene Transgenic mice with insertion of 200 kbp-long DNA fragments carrying a single bacterial artificial chromosome (BAC) clone Use of transposases (from either Tn5 or piggyBac) to insert transgenes into mouse chromosomes during ICSI. Highly increased efficiency of transgenesis (>60%)	Perry *et al.* 1999 Chan *et al.* 2000 Nagashima *et al.* 2003 Kurome *et al.* 2006 Moreira *et al.* 2004, 2007 Osada *et al.* 2005 Suganuma *et al.* 2005, Moisyadi *et al.* 2009, Wu *et al.* 2006
REMI	Amphibia Bovine, Chicken	Plasmid DNA integration with high efficiency in the genome of X.laevis embryos and tadpoles Efficient production of transgenic calves and chicken	Kroll and Amaya, 1999 Shemesh *et al.* 2000 Harel-Markowitz *et al.* 2009
LB-SMGT	Mouse, pig	Stable integration of transgenes and transmission to F1 and F2 progeny.	Chang *et al.* 2002
Liposomes	Rabbit adults Rat adults Chicken adults Horse embryos Bovine embryos *(continued from previous page)*	Transgene integration verified by PCR and Southern blot. Expression in embryos and in tissues of young rabbits. Transgene inheritance in F1 progeny Increased production of transgenic rats with liposome-protamine peptide-DNA complex Transgenic chicks, about 5% efficiency Efficiency of delivery, propagation and expression of foreign DNA significantly affected by the choice of transfection reagent and plasmid architecture.	Wang *et al.* 2001, 2003 Yonezawa *et al.* 2002 Yang *et al.* 2004 Ball *et al.* 2008 Hoelker *et al.* 2007
TMGT	Mouse adults Mouse adults Mouse adults Rat embryos and adults Mouse adults Fish adults Goat embryos Mouse adults Abalone	Plasmid injection into interstitial space of testis. Transgenic mice obtained by mating with normal females. Evidence of persistence of episomal structures. Transgene transmission from F0 to F1 and F2 generations. YFP-expressing transgenic spermatozoa generated by plasmid injection into seminiferous tubules. Transgenic mice produced by ICSI showed integration and transmission of the transgene to offspring Plasmid injected into vas deferens. Transgenic mice generated by normal mating Injection of DNA/liposome complexes with various tested liposomes. 80% positive embryos, but proportion decreases with development Injection of DNA/liposome complex into seminiferous tubules. Episomal transgene in offspring, found in the tail of young animals but not in adulthood Injection of DNA/liposome complex, transgenesis efficiency >80%. Transgene integration and transmission to offspring Direct plasmid DNA injection into testis. Transgene integration and transmission to F1 (41% efficiency) and F2 progeny (37% efficiency) DNA injection. 90% gene transfer efficiency in larvae and 60% in 1-year-old adults, respectively. Around 20% of G0 mature abalone contained the transgene in their gonads.	Sato *et al.* 1999a, 1999b Huang *et al.* 2000 Huguet and Esponda 2000 Yonezawa *et al.* 2001 Celebi *et al.* 2002 Lu *et al.* 2002 Li *et al.* 2005 He *et al.* 2006 Chen *et al.* 2006
Virus Mediated-Tr	Mouse adults Rat adults Rat Mouse, goat	Spermatogonial stem cells (SSCs) transduced with retrovirus injected into immature seminiferous tubules. After mating, transgene was in 2.8% of offspring, transmitted to progeny and expressed. Transgenic rats generated by immunodeficient founder mice	Kanatsu-Shinohara *et al.* 2004 Kanatsu-Shinohara *et al.* 2008

	adults Zebrafish	xenotransplanted with rat SSCs transduced *in vitro* with EGFP-expressing lentiviral vector	Ryu *et al.* 2007 Honaramooz *et al.* 2008 Kurita *et al.* 2004
		Rat SSCs transduced *in vitro* with EGFP-expressing lentiviral vector transplanted into recipient rat	
		testes, generating 6% stable transgenic offspring.	
		Transgenic mice and goats obtained with 10% efficiency from males transplanted with adeno-associated virus-transduced germ line stem cells	
		Retrovirus-infected SSCs from Zebrafish were differentiated *in vitro*, producing sperm cells that generated transgenic fish by IVF	
Electro poration	Bovine embryos Loach embryos Finfish, loach, shellfish larvae, adults Carp embryos and adult animals Carp adults Bovine embryos	65% transgenic larvae. Integration assessed by Southern blot. Fast growing GH-transgenic individuals Positive animals identified by PCR. Integration assessed by Southern blot. Mosaicism, coexistence of non-integrated and integrated foreign sequences. Correct expression of transgene	Gagné *et al.* 1991 Chen *et al.* 1997 Tsai *et al.* 1995, 1997, 2000 Sarangi *et al.* 1999 Venugopal *et al.* 2004 Rieth *et al.* 2000

INTRACYTOPLASMIC SPERM INJECTION-MEDIATED TRANSGENESIS (ICSI-TR)

This method, first described in mouse [25], stemmed from the combination of SMGT and the standard ICSI approach [26] widely used in assisted reproductive technology. In a first step, the sperm plasma membrane is disrupted by either freeze-thawing or Triton X-100 treatment. Membrane-disrupted spermatozoa are then briefly incubated with exogenous DNA. In the third and final step, the spermatozoa/DNA complex, unable to fertilise autonomously, is microinjected into metaphase II oocytes by ICSI, i) allowing the transgene to be incorporated into the embryonic genome *via* the DNA repair mechanism, and ii) activating the developmental programme [27]. This method yields a very high percentage (60 to 90%) of preimplantation embryos expressing EGFP or LacZ reporter genes. Following implantation of the ICSI-derived embryos into foster mothers, 20 to 45% of born offspring were found to be transgenic [25, 28, 29].

ICSI-Tr offers several advantages compared with the most popular method of zygote microinjection: it uses unfertilised oocytes, so the time window for manipulation is much more extended than the interval during which the two pronuclei are distinctly visible in zygotes. Furthermore, spermatozoa carrying the exogenous DNA can be injected anywhere in the cytoplasm of metaphase-arrested oocytes, and not necessarily in the male pronucleus. Transgene integration efficiencies of up to 100% have been reported. Possibly the greatest advantage offered by ICSI-Tr is the potential to insert very large DNA fragments (>200 kb) into the host genome [28, 30]. The large size can accomodate all required regulatory elements for proper transcription of the genomic gene, which guarantees faithful expression of the transgene.

Recently, an ICSI-Tr protocol with markedly high integration efficiency has been developed utilising transposon-encoded transposases [31]. Tn5 transposase mixed with isolated sperm heads, or with round spermatids, enhances the integration of transgenes into mouse chromosomes. In addition, a construct containing the piggyBac transposase was shown to be very efficient in generating transgenic mice [27, 32], improving the rate of mouse transgenesis to about 69 % of born animals [33].

It has been noticed that freeze-thawing or TritonX-100 treatment, required in the ICSI procedure to remove the plasma membrane, acrosome and tail before spermatozoa microinjection, can cause severe paternal chromosomal damage, which impairs further development of the generated embryos and offspring [34, 35]. Minimizing the damage to sperm chromatin is therefore expected to improve transgenesis.

A recent study reported a novel method for ICSI-Tr, in which spermatozoa are pretreated with a simple, inexpensive 10 mM NaOH medium before incubation with EGFP reporter plasmid [36]. The timing of incubation of sperm heads with exogenous DNA and the concentration of transgene are important parameters in determining the efficiency of this method; the highest rate of transgenesis was obtained using 2 ng/μL DNA incubated for less than 10 min with NaOH-pretreated spermatozoa. Under these conditions, about 35% of born pups were transgenics, a lower number of microinjected oocytes was required and efficient germline transmission was observed from F0 to F1 generation.

In conclusion, ICSI-Tr, incorporating transposons and NaOH-treated spermatozoa, has improved the efficiency of transgenic generation to a level approaching that of virus-mediated methods (see below), at least in the murine system. Furthermore, ICSI-Tr is thus far the only practical option when very large transgenes of >500 kb need inserting for correctly regulated gene expression.

RESTRICTION ENZYME-MEDIATED INTEGRATION (REMI)

In this technique, the exogenous plasmid DNA is first linearised with a restriction enzyme that generates sticky ends; the linearised plasmid is then transfected together with the restriction enzyme into sperm cells by lipofection. The lipofected restriction enzyme cleaves the genomic paternal DNA, creating sites that enable the exogenous DNA to integrate *via* its matching cohesive ends. REMI has been originally developed to insert exogenous DNA in nuclei of X. laevis sperm cells, which were then injected into unfertilised eggs to produce transgenic embryos [37]. A non-mosaic, high transgene expression rate was demonstrated following this procedure. More recently, transgenic frogs carrying one single integration event were obtained at high efficiency using a new approach employing the yeast transposase I-SceI [38].

The REMI protocol has been adapted to animal species of commercial interest, *e.g.* bovine, chicken and equine. In bovine transgenesis, REMI has been applied to both IVF and AI procedures [39]: in the former case, expression of EGFP reporter gene was detected in about 30% of morula-stage embryos, as determined by RT-PCR analysis of total RNA from blastomeres; in the latter, calves were produced by AI, all of which expressed the EGFP reporter gene. REMI has recently been extended to chicken in AI experiments with high rates of success [40] and *in vitro* to equine embryos [41]. On the whole, this method significantly improves the rate of integration of the transgene in the host genome.

LINKER-BASED SMGT (LB-SMGT)

An alternative protocol to increase the binding of exogenous DNA molecules to sperm cells is linker-based SMGT (LB-SMGT), first described in 2002 in mice and pigs [42]. In LB-SMGT, linearised DNA is bound in a non-covalent manner to a monoclonal antibody (mAb) that recognizes a specific antigen located on the surface of sperm membrane. DNA coupled with antibody offers the possibility to target selected sperm cells and facilitate the internalisation of the DNA complexes *via* receptor-mediated endocytosis. Spermatozoa of pig, mouse, chicken, cow, goat, sheep and humans have all been successfully tested and the results show that exposure to mAb-coupled DNA increases the uptake in sperms by 25-56% compared to controls incubated with no antibody or with non-specific mAb. Transgenic F0 pigs and mice have been obtained by LB-SMGT and a number of mosaic transgenic pigs displayed germ-line transmission. The mosaicism may be due to transgene integration at later stages of embryonic development. Germ-line transmission from F0 to the F1 generation occurred at an efficient rate, with 37.5% of pigs and 33% of mice testing positive for transgene presence. Transmission to F2 progeny was also demonstrated by FISH and Southern blot analysis, suggesting that the transgene was stably integrated in the host genome. Expression of the transgene was generally detected in a large proportion of positive animals (more than 60% of transgenic individuals).

TESTIS MEDIATED GENE TRANSFER (TMGT)

TMGT is an alternative and, in principle, a simplified variant of SMGT that does neither require IVF nor embryo transfer procedures. Foreign DNA sequences are directly injected into the interstitial space of the mammalian testis. Studies using either fluorescently labelled DNA [43] or trypan blue [44] suggest that foreign molecules introduced in the testis are transported to the cauda epididymis within 3-4 days after injection and are then taken up by epididymal spermatozoa, which deliver them to oocytes at fertilisation. Transgenic mice and rats have been generated using this procedure, the transgene was found in more than one copy per genome and was correctly expressed in individuals of both F0 and F1 generations [45]. Other works reported slightly different results, showing that the foreign DNA sequences were detected in the vast majority (80-100%) of the fetuses by PCR, whereas genomic Southern analysis failed to detect any specific signal in the same samples [46-48]; the exogenous sequences were however transmitted from F0 to the F1 and F2 generations [49]. These results suggest that foreign DNA sequences introduced into adult mouse testis are mosaically distributed in fetal tissues, have low abundance (*i.e.* below one copy/genome and below the resolution power of Southern blotting), are transmitted in a non-Mendelian fashion from one generation to the next and are progressively lost during development (extensively reviewed by Sato, 2005). Interestingly, these features are similar to those observed after direct sperm/DNA interaction, in which the foreign sequences remain as non-integrated extrachromosomal structures [11, 12, 48]. Similar results were obtained when the circular plasmid DNA/liposome complex was injected in mouse seminiferous tubules [50]. The DNA was extracted from the offspring of injected males mated to wild-type females and, under these conditions, PCR and Southern blot analysis showed again that the exogenous sequences were transmitted to the progeny but remained episomal.

Truly transgenic mice have instead been produced from males injected into vas deferens with exogenous DNA [51]. Microscopy analysis of fluorescent *in situ* hybridization (FISH)-treated spermatozoa collected from the injected vas deferens revealed the presence of the foreign gene in 70% of the cells. After mating, the transgene was indeed transmitted to the progeny but the percentage of genetically modified animals was significantly lower (7.5%) compared to the fraction of spermatozoa that contained the transgene.

The heterogeneity of the outcomes obtained with TMGT may depend on the different types of liposomes used in the association of foreign DNA with spermatozoa [47] and/or with the different sites of injection, *i.e.* interstitial space and seminiferous tubules of the testis or vas deferens.

TMGT has been successfully applied to fish [52] and shellfish [53] after injection of testis with either the liposome/transgene complex or with naked DNA, respectively. PCR analysis showed that, in both cases, the transgenes were abundantly propagated in larvae and maintained in the majority of adults; furthermore, Southern blot analysis revealed integration in the host genome of adult individuals. It is worth mentioning that the transgene used in experiments with the shellfish abalone was a growth hormone cDNA and the average shell size and body weight of transgenics were significantly larger than those of controls. Although not directly related to TMGT, it is worth mentioning in this context that transgenic silkworm pupae have recently been obtained by injecting plasmid DNA into the copulatory pouch of moths which were then free to copulate and oviposit [54].

VIRUS-MEDIATED GENE DELIVERY INTO MALE GERM CELLS

Viral vectors have historically been viewed as potential tools for animal transgenesis. Consistent with this, a recombinant lentiviral vector has been successfully used to develop a highly efficient protocol for transgenesis, ensuring the delivery of transgenes to one-cell mouse embryos [55]. Viral vectors are also used to deliver foreign genes to male germ cell precursors, aiming to produce "transgenic spermatozoa". In contrast with female germ cells, which stop proliferation before birth, spermatogonial stem cells (SSCs), *i.e.* the progenitors of all male germ cells, are subjected to continuous self-renewal [56]. In mice and rats, transgenesis protocols *via* transplantation of genetically modified male germ cells have made use of germ cells that were transduced *in vitro* with retroviral, lentiviral or adeno-associated viruses prior to transplantation [57-59]. These strategies yielded germ line transgenesis, but several drawbacks are present, including a modest transgenesis efficiency, the low fertility rate of the recipient animals following spermatogonial transplantation and the complex, highly demanding experimental procedure. Under this light, the practical value of this approach is limited and especially prevents the large scale application of the technique for animal transgenesis.

In vivo transduction of male germ stem cells has emerged as a more practical alternative for the production of transgenic offspring. SSC can be transduced by microinjecting a retroviral vector into immature seminiferous tubules and transgenic mice are generated after mating with wild-type females [60]. A similar approach has been used by the same group to produce knock-out mice [61]. A further development of this procedure takes advantage of xenogeneic transplantation, in which rat SSCs were transduced *in vitro* with EGFP-expressing lentiviral vector and then transplanted into the testes of immunodeficient mice. The transduced rat SSCs produced EGFP-expressing spermatogenic cells that were used to produce transgenic rats, which stably transmitted the transgene to the next generation [62]. The microinjection of viral vectors *in vivo* into the testis circumvents the drawbacks associated with the *in vitro* transduction approach and provides an improved alternative protocol for SMGT.

SPERM-MEDIATED EMBRYONIC GENE THERAPY

After the finding that spermatozoa can be used as vectors of foreign genetic information, the focus on these cells has shifted from the strict reproductive field to that of genome manipulation. As summarised above, the genetic and phenotypic traits generated by SMGT after direct sperm/DNA interaction (direct SMGT) include a lack of integration of the exogenous sequences in the host genome and their mosaic persistence as extrachromosomal structures. These features represent limitations from the perspective of generating new transgenic animal strains, but are instead of high value in a gene therapy perspective. As a matter of fact, the ability of mature spermatozoa to generate and propagate transcriptionally competent 'retrogenes' (see chapter 12) that remain non-integrated and therefore do not interrupt the integrity of the recipient genome is a highly valuable parameter that can be usefully applied in gene therapy. On these grounds, the new perspective of sperm mediated embryonic gene therapy (SMEGT) may be envisaged. Direct SMEGT offers several advantages:

- Using direct SMEGT is an alternative to using viruses as delivery vectors of foreign genes in germ stem cells. At this time, viral vectors are considered to be the most efficient available transfection tools; on the other hand, however, insertional mutagenesis – *i.e.* the disruption of human genes, or the alteration of their normal expression - is a major cause of concern for human health [63], limiting their use in human therapy.

- Direct SMEGT would also circumvent the similar safety concerns raised by ICSI-Tr [25], which has been proposed as a potential alternative approach to human gene therapy [reviewed in [64]. In this case, in addition to the concern for uncontrolled integration, another issue is represented by the fact that such experiments cause the death of 80–90% of microinjected oocytes and embryos. Such a high rate of embryo letality may be acceptable in mice transgenesis, but seems hardly acceptable in human gene therapy.

- A further advantage offered by direct SMEGT is that sperm cells can deliver to embryos exogenous DNA of apparently unlimited size, thus allowing a well-regulated expression of the foreign gene, or set of genes, introduced in their "natural" regulatory DNA context. This is a significant advantage in comparison to viral delivery, which has a limited size of insertable DNA, but not to ICSI-Tr.

- Finally, direct SMEGT allows the introduction and expression of novel functional traits as early as the zygote stage: this can provide an opportunity of treatment before the damage or defects associated with the specific pathology become manifest.

On these grounds, direct SMEGT can provide a potential method for embryonic gene therapy with several advantages over other methods, especially those using viral vectors.

CONCLUSIONS

In the light of the results reported here, the binding and internalisation of exogenous DNA within sperm cells have disclosed a highly sophisticated network of functions in mature sperm cells. In SMGT studies, spermatozoa emerge as carriers of a complex machinery that determines the fate of the exogenous DNA molecules, *i.e.* either integration into the host genome *via* repair mechanism and Mendelian transmission to the progeny, or, more commonly, lack of integration, reverse transcription and transmission of phenotypic traits unlinked to chromosomes and propagated as non-Mendelian characters (see also Chapter 12 in this book). Both alternatives have far-reaching implications for the fate of the recipient embryos, fetuses or born individuals. Modulating SMGT conditions provides the opportunity to activate preferentially either one or the other mechanism. By achieving full experimental control over this mechanism, it becomes possible to generate new genetic or epigenetic modifications in the recipient organisms. In these perspectives, spermatozoa play a crucial role in biotechnology. If confirmed that these mechanisms can be activated in nature - and at this stage we have no reason to doubt that this is indeed the case - then their relevant implications for health and evolution should be seriously considered.

REFERENCES

[1]   Miller D, Ostermeier GC, Krawetz SA. The controversy, potential and roles of spermatozoal RNA Trends Mol Med 2005; 11: 156-63.

[2]   Krawetz SA. Paternal contribution: new insights and future challenges. Nat Rev Genet 2005; 6: 633-42.

[3]   Dadoune JP. Spermatozoal RNA: what about their functions? Microsc Res Tech 2009; 72: 536-51.

[4]   Ostermeier GC, Dix DJ, Miller D, *et al.* Spermatozoal RNA profiles of normal fertile men. Lancet 2002; 360: 772-7.

[5]   Ostermeier GC, Goodrich RJ, Moldenhauer JS, *et al.* A suite of novel human spermatozoal RNAs. J Androl 2005; 26: 70-74.

[6]   Wagner KD, Wagner N, Ghanbarian H, *et al.* RNA induction and inheritance of epigenetic cardiac hypertrophy in the mouse. Dev Cell 2008; 14: 962-69.

[7]   Pittoggi C, Magnano AR, Sciamanna I, *et al.* Specific localization of transcription factors in the chromatin of mouse mature spermatozoa. Mol Reprod Dev 2001; 60: 97-106.

[8]   Gur Y, Breitbart H. Mammalian sperm translate nuclear-encoded proteins by mitochondrial-type ribosomes. Genes Dev 2006; 20: 411-16.

[9]   Ostermeier GC, Miller D, Huntriss JD, *et al.* Delivering spermatozoan RNA to the oocyte. Nature 2004; 429: 154.

[10]   Rassoulzadegan M, Grandjean V, Gounon P, *et al.* RNA-mediated non-mendelian inheritance of an epigenetic change in the mouse. Nature 2006; 441: 469-74.

[11]   Spadafora C. Sperm-mediated "reverse" gene transfer: a role of reverse transcriptase in the generation of new genetic information. Hum Reprod 2008; 23: 735-40.

[12]   Sciamanna I, Vitullo P, Curatolo A, *et al.* Retrotransposons, reverse transcriptase and the genesis of new genetic information. Gene 2009; 448: 180-86.

[13]   Spadafora C. Sperm cells and foreign DNA: a controversial relation. BioEssays 1998; 20: 955-64.

[14]   Smith K, Spadafora C. Sperm-mediated gene transfer: applications and implications. Bioessays 2005; 27: 551-62.

[15]   Niu Y, Liang S. Progress in gene transfer by germ cells in mammals. J. Genet. Genomics 2008; 35: 701-14.

[16]   Lavitrano M, Camaioni A, Fazio V, *et al.* Sperm cells as vectors for introducing foreign DNA into eggs: genetic transformation of mice. Cell 1989; 57: 717-23.

[17]   Kim TS, Lee SH, Gang GT, *et al.* Exogenous DNA uptake of boar spermatozoa by a magnetic nanoparticle vector system. Reprod Domest Anim 2010; 45: 201-06.

[18]   Arezzo F. Sea-Urchin Sperm as a vector of foreign genetic information. Cell Biol Int Rep 1989; 13: 391-404.

[19]   Sciamanna I, Piccoli S, Barberi L, *et al.* DNA dose and sequence dependence in sperm-mediated gene transfer. Mol Reprod Dev 2000; 56: 301-05.

[20]   Sciamanna I, Barberi L, Martire A, *et al.* Sperm endogenous reverse transcriptase as mediator of new genetic information. Biochem Bioph Res Comm 2003; 312, 1039-46.

[21]   Pittoggi C, Beraldi R, Sciamanna I, *et al.* Generation of biologically active retro-genes upon interaction of mouse spermatozoa with exogenous DNA. Mol Reprod Dev 2006; 73: 1239-46.

[22]   Zoraqi G, Spadafora C. Integration of foreign DNA sequences into mouse sperm genome. DNA Cell Biol 1997; 16, 291-00.

[23]   Lavitrano M, Bacci ML, Forni M, *et al.* Efficient production by sperm-mediated gene transfer of human decay accelerating factor (hDAF) transgenic pigs for xenotransplantation. Proc Natl Acad Sci USA 2002; 99: 14230-35.

[24]   Lavitrano M, Forni M, Bacci ML, *et al.* Sperm mediated gene transfer in pig: selection of donor boars and optimization of DNA uptake. Mol Reprod Dev 2003; 64: 284-91.

[25] Perry AC, Wakayama T, Kishikawa H, *et al.* Mammalian transgenesis by intracytoplasmic sperm injection. Science 1999; 284: 1180-83.

[26] Kimura Y, Yanagimachi R. Intracytoplasmic sperm injection in the mouse. Biol Reprod 1995; 52: 709-20.

[27] Moisyadi S, Kaminski JM, Yanagimachi R. Use of intracytoplasmic sperm injection (ICSI) to generate transgenic animals. Comp Immunol, Microb Infect Dis 2009; 32: 47-60.

[28] Moreira PN, Giraldo P, Cozar P, *et al.* Efficient generation of transgenic mice with intact yeast artificial chromosomes by intracytoplasmic sperm injection. Biol Reprod 2004; 71: 1943-47.

[29] Hirabayashi M, Hochi S. Generation of transgenic rats by ooplasmic injection of sperm cells exposed to exogenous DNA. Methods Mol Biol 2010; 597: 127-36.

[30] Osada T, Toyoda A, Moisyadi S, *et al.* Production of inbred and hybrid transgenic mice carrying large (>200kb) foreign DNA fragments by intracytoplasmic sperm injection. Mol Reprod Dev 2005; 72: 329-35.

[31] Suganuma R, Pelczar P, Spetz JF, *et al.* Tn5 transposase-mediated mouse transgenesis. Biol Reprod 2005; 73: 1157-63.

[32] Wu SC, Meir YJ, Coates CJ, *et al.* piggyBac is a flexible and highly active transposon as compared to Sleeping Beauty, Tol2, and Mos1 in mammalian cells. Proc Natl Acad Sci USA 2006; 103: 15008-13.

[33] Shinohara ET, Kaminski JM, Segal DJ, *et al.* Active integration: new strategies for transgenesis. Transgenic Res 2007; 16: 333-39.

[34] Yamauchi Y, Doe B, Ajduk A, *et al.* Genomic DNA damage in mouse transgenesis. Biol Reprod 2007; 77: 803-12.

[35] Yamagata K, Suetsugu R, Wakayama T. Assessment of chromosomal integrity using a novel live-cell imaging technique in mouse embryos produced by intracytoplasmic sperm injection. Hum Reprod 2009; 24: 2490-99.

[36] Li C, Mizutani E, Ono T, *et al.* An efficient method for generating transgenic mice using NaOH-Treated spermatozoa. Biol Reprod 2009; 82: 331-40.

[37] Kroll K, Amaya E. Transgenic Xenopus embryos from sperm nuclear transplantations reveal FGF signaling requirements during gastrulation. Development 1996; 122: 3173-83.

[38] L'hostis-Guidet A, Recher G, Guillet B, *et al.* Generation of stable Xenopus laevis transgenic lines expressing a transgene controlled by weak promoters. Transgenic Res 2009; 18: 815-27.

[39] Shemesh M, Gurevich M, Harel-Markowitz E, *et al.* Gene integration into bovine sperm genome and its expression intransgenic offspring. Mol Reprod Dev 2000; 56: 306-08.

[40] Harel-Markowitz E, Gurevich M, Shore LS, *et al.* Use of sperm plasmid DNA lipofection combined with REMI (Restriction Enzyme-Mediated Insertion) for production of transgenic chickens expressing eGFP (Enhanced Green Fluorescent Protein) or human folliclestimulating hormone. Biol Reprod 2009; 80: 1046-52.

[41] Ball BA, Sabeur K, Allen WR. Liposome-mediated uptake of exogenous DNA by equine spermatozoa and applications in sperm-mediated gene transfer. Equine Vet J 2008; 40: 76-82.

[42] Chang K, Qian J, Jiang MS, *et al.* Effective generation of transgenic pigs and mice by linker based sperm-mediated gene transfer. BMC Biotechnol 2002; 2: 5-17.

[43] Chang KT, Ikede A, Hayashi K, *et al.* Possible mechanism for the testis mediated gene transfer as a new method for producing transgenic animals. J Reprod Dev 1999a; 45: 37-42.

[44] Sato M, Ishikawa A, Kimura M. Direct injection of foreign DNA into mouse testis as a possible *in vivo* gene transfer system *via* epididymal spermatozoa. Mol Reprod Dev 2002; 61: 121-35.

[45] Chang KT, Ikede A, Hayashi K, *et al.* Production of transgenic rats and mice by the testismediated gene transfer. J Reprod Dev 1999b; 45: 29-36.

[46] Sato M, Gotoh, K, Kimura M. Sperm-mediated gene transfer by direct injection of foreign DNA into mouse testis. Transgenics 1999; 2: 357-69.

[47] Yonezawa T, Furuhata Y, Hirabayashi K, *et al.* Detection of transgene in progeny at different developmental stages following testis-mediated gene transfer. Mol Reprod Dev 2001; 60: 196-201.

[48] Sato M, Nakamura S. A novel gene transmission pattern of exogenous DNA in offspring obtained after testis-mediated gene transfer (TMGT). Transgenics 2004; 4: 121-35.

[49] Sato M, Yabuki K, Watanabe T, *et al.* Testis-mediated gene transfer (TMGT) in mice: succesful transmission of introduced DNA from F0 to F2 generations. Transgenics 1999; 3: 11-22.

[50] Celebi C, Auvray P, Benvegnu T, *et al.* Transient transmission of a transgene in mouse offspring following *in vivo* transfection of male germ cells. Mol. Reprod Dev 2002; 62: 477-82.

[51] Huguet E, Esponda P. Generation of genetically modified mice by spermatozoa transfection *in vivo*: preliminary results. Mol Reprod Dev 2000; 56: 243-47.

[52]   Lu JK, Fu BH, Wu JL, *et al.* Production of transgenic silver sea bream (Sparus sarba) by different gene transfer methods. Mar Biotechnol 2002; 4: 328-37.

[53]   Chen HL, Yang HS, Huang R, *et al.* Transfer of a foreign gene to Japanese abalone (Haliotis diversicolor supertexta) by direct testis-injection. Aquaculture 2006; 253: 249-58.

[54]   Li Y, Cao G, Chen H, *et al.* Expression of the hGM-CSF in the silk glands of germline of gene-targeted silkworm Biochem Biophys Res Comm 2010; 391: 1427-31.

[55]   Lois C, Hong EJ, Pease S, *et al.* Germline transmission and tissue-specific expression of transgenes delivered by lentiviral vectors. Science. 2002; 295: 868-72.

[56]   de Rooji DG. The spermatogonial stem cell niche. Microsc Res Tech 2009; 72: 580-85.

[57]   Nagano M, Brinster CJ, Orwig K, *et al.* Transgenic mice produced by retroviral transduction of male germ-line stem cells. Proc Natl Acad Sci USA 2001; 98: 13090-95.

[58]   Nagano M, Watson D.J, Ryu BY, *et al.* Lentiviral vector transduction of male germ line stem cells in mice. FEBS Lett. 2002; 524: 111-15.

[59]   Honaramooz A, Megee S, Zeng W, *et al.* Adenoassociated virus (AAV)-mediated transduction of male germ line stem cells results in transgene transmission after germ cell transplantation. FASEB J 2008; 22: 374382.

[60]   Kanatsu-Shinohara M, Toyokuni S, *et al.* Transgenic Mice Produced by Retroviral Transduction of Male Germ Line Stem Cells *In vivo.* Biol Reprod 2004; 71: 1202-07.

[61]   Kanatsu-Shinohara M, Ikawa M, Takehashi M, *et al.* Production of knockout mice by random and targeted mutagenesis in spermatogonial stem cells. Proc Natl Acad Sci USA 2006; 103: 8018-23.

[62]   Kanatsu-Shinohara M, Kato M, Takehashi M, *et al.* Production of transgenic rats *via* lentiviral transduction and xenogeneic transplantation of spermatogonial stem cells. Biol Reprod 2008; 79: 1121-28.

[63]   Nair V. Retrovirus-induced oncogenesis and safety of retroviral vectors. Curr Opin Mol Ther 2008; 10: 431-88.

[64]   Navarro J, Risco R, Toschi M, *et al.* Gene therapy and intracytoplasmatic sperm injection (ICSI) – a review. Placenta 2008; 29: 193-99.

[65]   Lavitrano M, Maione B, Forte E, *et al.* The interaction of sperm cells with exogenous DNA: a role of CD4 and major histocompatibility complex class II molecules. Exp Cell Res 1997; 233: 56-62.

[66]   Maione B, Pittoggi C, Achene L, *et al.* Activation of endogenous nucleases in mature sperm cells upon interaction with exogenous DNA. DNA Cell Biol 1997; 16: 1087-97.

[67]   Francolini M, Lavitrano M, Lora Lamia C, *et al.* Evidence for nuclear internalisation of exogenous DNA into mammalian sperm cells. Mol Reprod Dev 1993; 34: 133-39.

CHAPTER 3

Evolutionary Implications of SMGT

Kevin R. Smith[*]

Abertay University, United Kingdom

Abstract: Strong natural barriers exist against SMGT. However, such barriers are unlikely to be absolutely inviolable. If sperm cells can indeed act as vectors for exogenous DNA, it follows that the genome of sexually reproducing animals may be subject to alteration by exogenous DNA sequences carried by sperm cells. At present there are insufficient data to permit quantification of the rate at which SMGT may occur in nature. Nevertheless, the implications of such 'natural' SMGT are significant, and include evolutionary effects on the mammalian genome and pathologies in humans from *de novo* mutations.

Keywords: Concatemeristion, *De novo* mutation, Gene knockout, Genome evolution, Homologous recombination, Mammalian genome, Natural SMGT, Random integration, Repetitive sequence, Retrotransposition, Tandem array.

INTRODUCTION AND BACKGROUND

As discussed in chapter 1, sperm cells may be able to attach and transport exogenous DNA molecules (exogenes) into the oocyte at fertilisation. If sperm cells are able to act in this way, the evolved structure of the human genome (and that of other sexually reproducing animals) should reflect this phenomenon. Further, human embryos are potentially at risk from mutations caused by sperm acting as exogene vectors.

Dating from the late 80s, many reports of the successful *in vitro* uptake of exogene constructs (transgenes) by animal sperm cells have been published [1, 2], with many of these reports being discussed in this book. A majority of the published reports provide evidence of post-fertilisation transfer and maintenance of transgenes. Several studies report the subsequent generation of viable progeny animals, the cells of which contain transgene DNA sequences. While several published studies have used 'augmentation' techniques (for example electroporation or liposomes) to 'force' sperm to capture exogenes, the original methodology is very straightforward: prior to IVF or AI, 'washed' sperm cells are simply incubated in a DNA-containing solution [1, 2]. This 'autouptake' approach is still employed by some SMGT experimenters. As a potential tool for genetically manipulating animals, sperm-mediated gene transfer (SMGT) has the advantages of simplicity and cost-effectiveness, in contrast with more established methods of transgenesis such as pronuclear micrinjection.

However, despite the above successes and regardless of its potential utility, SMGT has not become established as a reliable form of genetic manipulation. Concerted attempts to utilise SMGT have often produced negative results. The most notable example of such a failure is to be found in the collated results of several independent research groups: of 890 mice analysed, not a single animal contained transgene DNA [3].

Indeed, following the original reports of successful autouptake -based SMGT, followed by the failure to replicate these findings, some biologists were led to express skepticism over the fundamental biological basis for SMGT [4, 5]. Such skepticism is posited on the assumption that major evolutionary chaos would result if sperm cells were able to act as exogene vectors. Given that the reproductive tracts contain 'free' DNA molecules (originating from natural cell death and breakage), it seems reasonable to expect sperm cells to be highly resistant to the risk of picking up such molecules.

*****Address correspondence to Kevin R. Smith:** School of Contemporary Sciences, Abertay University, Dundee, DD1 1HG, United Kingdom; Tel: 44 (0)1382 308664; E-mail: k.smith@tay.ac.uk

Nevertheless, there now exists a well established body of empirical data showing that sperm cells are able, under particular experimental circumstances, to interact with and carry exogenes [6]. Furthermore, isolated reports of the successful use of SMGT for genetic modification continue to be published. Recent examples include the generation of several transgenic pigs, rabbits and goats following insemination with sperm cells preincubated with transgene DNA [7-9].

There are two possible ways to make sense of the above experimental and theoretical considerations. The first possible explanation is that SMGT is fundamentally unattainable. If so, the empirical evidence in support of SMGT must be faulty. For example, perhaps sperm can associate with exogenous DNA but cannot convey the DNA into the oocyte; and transgene sequences may have been erroneously identified in tissue samples, perhaps due to DNA contamination affecting sensitive detection methods such as PCR. This scenario is certainly not impossible: scientific research contains several examples of theory being misled by mistaken data. Indeed, early reports of SMGT were compared with the (then contemporary) claims of "cold fusion" in physics [4]. By contrast, the second possible explanation is that SMGT is viable, and that the claims of experimental success were not made in error. If this explanation is correct, it follows that 'horizontal inheritance' *via* SMGT must be a natural occurrence in higher animals. Given that evolutionary chaos is not observed in nature, this 'natural' SMGT presumably occurs at a low frequency.

It is beyond the scope of the present discussion to attempt to decide between these competing explanations. Indeed, much more research is needed before a consensual resolution is achieved. Meanwhile, however, it is valuable to explore the wider biomedical implications arising from the possibility that SMGT does occur in nature. Specifically, it is hypothesized that SMGT may have played a role in the evolution of the mammalian genome, and that SMGT may cause mutations in human embryos.

DNA UPTAKE IN THE REPRODUCTIVE TRACTS

Exogene uptake by sperm is arguably most likely within the female tract. During maturation, sperm become coated with glycoproteins. From *in vitro* studies, it appears that exogene uptake by sperm is strongly antagonised by inhibitory glycoproteins (see Chapter 1). Sperm glycoproteins remain in place during ejaculation, and are only removed during capacitation, a process that occurs in the oviduct. Thus, the oviduct seems to be the most likely site for exogene uptake.

It is well accepted that, due to natural cell death and breakage, the oviduct must contain 'free' DNA molecules. However, there exists a dearth of data on the quantities of DNA present. This lack of knowledge makes it very difficult to use data from *in vitro* sperm-exogene studies to speculate in quantitative terms on the rate of exogene uptake *in vivo*. Nevertheless, there is much that can be said concerning what may and may not be possible within the oviduct.

In terms of the quantity of exogenes per sperm cell, studies across a range of species suggest three consistent features: (a) exogenous DNA quantity is positively correlated with the amount of DNA taken up by sperm; (b) internalised exogenous DNA, when present above a threshold amount (10-30 ng/106 sperm cells), triggers sperm cell death; and (c) exogenous DNA carried by sperm exerts a dose-dependent toxic effect on embryos [6, 10-12]. Thus, the chances of any sperm cell becoming a vector may be minimised by the need for dissolved DNA molecules to be present at neither too high nor too low a concentration in the oviduct.

An optimum exogene concentration appears to be in the region of 1 ng/106 sperm cells [11]. This translates into (very approximately) one cellular genome per fifty sperm cells. This is a fairly small amount of DNA, considering the large numbers of cells lining the oviduct. Given the present unavailability of data on exogene levels, it is not possible to be certain that adequate quantities of DNA are to be encountered in the oviduct. However, to the present author at least, it would be surprising if levels of around 1 ng DNA/106 sperm never occurred. Even if the 'basal' exogene level was too low, this level would not remain static. Damage or infection within the oviduct should result in more genomic DNA being released, due to increased apoptosis, and infective agents themselves may add exogene molecules when they die. In the other direction, exogene molecules would be broken-down by nuclease enzymes. Speculatively, windows of opportunity for sperm-exogene uptake may occur in nature, when exogene levels transiently optimise through chance events such as infection within the oviduct.

Note that it remains possible that the epididymis/vas deferens - as opposed to the oviduct - may serve as a site for exogene uptake. Recent unpublished data suggest that unrecognized anaerobes are more common than previously though in this location (A.A. Kiessling, personal communication), and thus might lead to a greater quantity of exogenous DNA being present. Sperm would encounter these anatomic components of the tract before becoming coated with inhibitory seminal vesicle proteins. Thus, it remains possible that the male tract may also be able to function as a site for exogene uptake by sperm.

The length of exogene molecules may have an effect on the probability of sperm-exogene uptake. It seems reasonable to expect smaller DNA molecules to be transferred with greater ease, considering the physical and energetic requirements that would be needed to get large exogenes into sperm cells. (Indeed, a general feature of all 'biological vector' forms of gene transfer is that there are limits on the maximum size of transgene able to be efficiently carried.) However, the little empirical data available on this aspect is contradictory. One study found uptake efficiency to be inversely proportionate to the length of the molecules within the range 1.25 to 2.5 kb [13], whereas another study found no relationship between uptake and length within the range 0.125 to 23 kb [14]. In the oviduct, one would expect to find a wide range of DNA sizes, generated by nuclease breakdown of chromosomal DNA. If there is an optimum size range for sperm-exogene uptake, the fact that only a proportion of the available DNA would be within this range would be another factor acting to reduce the frequency of sperm-exogene uptake in nature.

Exogene sequence structure is a final factor in the sperm-exogene uptake probability picture. It is noteworthy that several studies report that DNA sequences differ (under the same experimental conditions) in terms of the avidity with which they bind to sperm, suggesting that the primary structure of exogenes is not irrelevant to sperm-exogene uptake [11, 15]. Insufficient data presently exists to draw firm conclusions concerning which exogene sequences may be the most likely candidates for uptake by sperm. However, on the assumption that not all sequences within a fragmented genome are contenders for uptake by sperm cells, the mix of exogene sequence structures *in vivo* is expected to reduce the rate of sperm-exogene uptake.

If the empirical findings are correct and sperm cells can act as vectors experimentally, it is reasonable to conclude that sperm in the reproductive tracts should be expected to behave likewise in nature. However, the frequency of sperm-exogene uptake is likely to be strongly minimised by several factors, including the presence of inhibitory glycoproteins and the prevalence of non-optimum conditions in terms of exogene quantity, length, and primary structure.

EXOGENE INTEGRATION INTO THE HOST GENOME

Five possible fates await exogenes following their uptake by sperm cells. The first possibility is nuclease destruction, in which case modification of the host genome will not occur. It is well known that cell nuclei contain nuclease enzymes, presumably as a form of defence against genetic parasites such as retroviruses. Indeed, in transgenesis it is usual for integration to fail in a majority (60-90%) of successfully transfected oocytes, a phenomenon generally ascribed to nuclease activity [16]. Nuclease destruction is probably more likely within the oocyte rather than the sperm cell, given the generally inert nature of sperm cells.

The second possible fate for exogenes is integration into the sperm nucleus, prior to fertilisation. In view of the traditionally envisioned nature of the sperm cell nucleus as a metabolically inert entity, this form of integration would be viewed as surprising by many biologists. Nevertheless, early empirical claims to this effect were made, accompanied by some evidence to suggest that such integration may involve preferential sites within the sperm genome [17, 18]. Moreover, this work has been expanded upon in subsequent years, leading to the radical new theory that sperm are not in fact metabolically inert but rather can permit transgenes to integrate into the sperm genome, as explored in Chapter 12. However, such genomic integration appears to occur almost exclusively under experimental circumstances in which the sperm nucleus is directly accessed, without the transgene having to negotiate the nuclear membrane. Thus, although the occurrence in nature of sperm nuclear integration certainly cannot be ignored as a possibility, it seems likely that sperm cells sufficiently damaged to permit exogenes to bypass their membranes would inevitably not be viable for fertilisation.

The third possible fate is episomal maintenance in the cells of the resulting animal. Assuming that exogenes do not integrate into the sperm nucleus, episomal maintenance is conceptually possible. Episomally occurring transgenes have been reported following conventional transgenesis [19-21] and SMGT [22, 23]; indeed our current models of SMGT envisage an episomal route as being a key feature of SMGT in respect of transgenes taken up by intact spermatozoa, as eloborated in Chapter 12. However, specialised (*e.g.* viral) sequences are necessary for episomal maintenance to persist; otherwise, episomal sequences suffer from copy number instability and are lost fairly rapidly. On the basis that most exogenes in the oviduct are derived from fragmented chromosomal DNA, it is unlikely that the necessary sequences for episomal maintenance would be present in exogenes taken up by sperm cells in nature. Thus, it seems unlikely that episomally maintained exogenes delivered by SMGT in nature would have any significant evolutionary implications.

Genomic integration within the oocyte is a fourth possibility. If exogenes are experimentally brought into the nucleus of any animal cell (including the oocyte), random integration apparently occurs by a common pathway, regardless of the means of DNA delivery (*e.g.* co-precipitation, electroporation, microinjection) [24, 25]. Thus, it is attractive to speculate that exogenes delivered by the sperm cell may integrate *via* this common pathway. Certain signature features of post-fertilisation integration need to be identified in order to substantiate this speculation. However, little of the necessary analytical research (restriction analysis, sequencing) has been conducted on genetically modified animals resulting from SMGT. Thus it is not known whether exogene integration takes place within the oocyte.

The final possible fate for exogenes is non-random genomic integration, in which homologous recombination (HR) delivers the exogene to a homologous genomic site. Except where an exogene has originated from a foreign (*e.g.* bacterial) genome, homology between exogene and host genome is to be expected, given that exogenes in the oviduct are broken pieces of cellular DNA. However, non-random integration events in oocytes are outnumbered by random integration events by a factor of around 500 [26], therefore the non-random route should be expected to be especially rare.

CONSEQUENCES OF EXOGENE INTEGRATION

The simplest outcome of SMGT would be integration of a non-coding sequence into a functionless region of the genome, at a distance from any endogenous genes. In such cases, no phenotypic effects, and hence no medical or evolutionary consequences, would result. Given the well-known predominance of apparently functionless sequences within the mammalian genome, it follows that the majority of SMGT outcomes should have no immediate biological consequences.

Free from selective pressure, these 'silent' exogenes would remain in place within the genome as 'junk' DNA. In transgenesis, the process of concatemerisation (a part of the previously mentioned 'common pathway') frequently generates multiple transgene copies arranged in direct tandem array at the site of insertion [24, 27]. It is well known that the mammalian genome contains many tracts of repetitive sequences, many of which are present as direct tandem arrays. Thus, the possibility exists that, by contributing some of these repetitive sequence tracts, SMGT may have played a role in the 'junk' architecture of the mammalian genome.

It is doubtful that entire, intact genes could be transferred by SMGT, considering (a) the 'broken-down' origin of exogenes, and (b) the probable existence of limits on the length of exogenes in SMGT. In the unlikely event of an intact gene being transferred, a gain-of-function effect would be possible, assuming that the necessary control sequences were included in the exogene. However, significant gene expression would not be a foregone conclusion: transgene expression can vary by as much as 1000-fold depending on the site of genomic integration [28]. A more probable occurrence would be integration of gene fragments. This would be of potential evolutionary importance in cases where a fragment inserted close to an endogenous gene, or within an intron. In eukaryotes, the movement of coding sequences into proximity is an important recent source of new genes [29]. The primary mechanism for this movement may be retrotransposition [30]. SMGT represents an additional possible form of 'exon shuffling', and hence a potential source of new genes, in evolution.

As discussed previously, non-random genomic integration (in which HR delivers the exogene to a homologous genomic site) is expected to be very rare compared with random integration. However, if random integration *via* SMGT does occur in nature, it follows that non-random events will occasionally occur also.

Considered as a general phenomenon, non-random integration can proceed in either a conservative or non-conservative fashion [31]. In the context of SMGT, conservative recombination would result in endogenous sequences being overwritten by exogene sequences. Where an endogenous sequence was functional, biological effects could arise following sequence alteration. In conservative recombination, such alterations could be caused (a) by non-identity between the recombining sequences, and (b) *via* 'gene conversion' (an inherent part of the HR mechanism).

Non-conservative recombination would place the exogenous sequence close to its endogenous counterpart. In cases involving coding sequences, the result would be intragenic exon duplication. It is well known that many eukaryotic genes show evidence of intragenic duplications, and it is generally believed that such duplications have been important in the evolution of new genes. A variety of mechanisms successfully explain intragenic exon duplication, such as unequal crossover or unequal sister chromatid exchange. It is possible that SMGT operates as an additional mechanism in the duplication of intragenic DNA during evolution.

Finally, it is important to consider the simplest mode of biologically significant exogene integration: the random insertion of non-coding exogene DNA into a functional part of an endogenous gene. The almost inevitable consequence would be to eliminate (or 'knockout') the activity of the endogenous gene. In contrast to the other possible ways in which SMGT might alter the genome, gene knockout would have less evolutionary significance but considerably greater potential medical importance. In an evolutionary context, genes rendered functionless by insertional mutagenesis would be of minimal importance, because they would tend to be eliminated by natural selection. In medical terms, however, gene knockout by SMGT is potentially important as a form of *de novo* mutation in human embryos.

CONCLUSIONS

Several reports of the successful use of SMGT have been published, but attempts to repeat these experiments have not always met with success. Thus, SMGT has not been unequivocally accepted by biologists. It may be that nature has engineered an absolute barrier against sperm cells acting as exogene vectors. Certainly, if SMGT occurred at a high frequency in nature, the resultant chaotic inheritance patterns would have been detected in the early years of genetics.

An alternative explanation is that nature has erected formidable barriers against SMGT, but that these barriers are not absolute. On this view, the inconsistent nature of SMGT experimental outcomes is readily explained: if there are powerful natural barriers against SMGT, it follows that SMGT successes represent unusual cases in which the barriers have failed. Such breaches may have resulted from the presence of some unknown critical factor(s) in the successful experiments.

Absolute barriers against deleterious occurrences may exist, but they certainly are not the norm in biological systems. For example, polyspermy is deleterious, and there are powerful and well elucidated barriers against its occurrence, yet polyspermy happens in around 1% of natural human conceptions. And there are strong barriers against germline mutations – yet evolution is fuelled by such mutations, and *de novo* mutations do of course cause many illnesses. Thus it seems sensible to avoid a dogmatic supposition that SMGT must never occur.

If SMGT does occur in nature, it is impossible to estimate its frequency, given the present lack of data. If the rate of SMGT is very low, it may have only a slight evolutionary importance and virtually no medical importance. This might be somewhat analogous to DNA transposition, which has had a role in the architecture of the mammalian genome but only very rarely causes disease [32]. On the other hand, SMGT might turn out to occur at a higher rate, with concomitantly greater medical significance. This certainly cannot be ruled out: the human genome undoubtedly contains many future surprises. Trinucleotide repeat expansions (TREs) represent a salutary example. TREs are a major type of medically important mutation, yet they came as a surprise to geneticists upon their discovery in the early Nineties [33]. The sequence structure of the human genome undoubtedly contains many enigmas, and the origin of some genes appears to be driven at least in part by horizontal gene transfer, involving transposable elements [34, 35]. It is possible that ancestral sperm may have severed as vehicles for such transfers, *via* uptake of exogenous sequences in nature.

Clearly, there is a need for more research into SMGT per se. Meanwhile, as the human and animal genome projects unfold, it would be wise for geneticists to be on the lookout for telltale signs of SMGT in the sequence data. Attention should also be paid to the potential evolutionary implications of SMGT. Finally, clinical geneticists should be aware of the possibility that SMGT may represent a new form of medically relevant mutation.

ACKNOWLEDGEMENT

This chapter is based on an updated adaptation of a previously published paper by the author [36].

REFERENCES

[1]   Smith KR. Sperm cell mediated transgenesis: a review. Anim Biotechnol 1999; 10: 1-13.

[2]   Smith K, Spadafora C. Sperm-mediated gene transfer: applications and implications. BioEssays 2005; 27: 551-62.

[3]   Brinster RL, Sandgren EP, Behringer RR, *et al.* No simple solution for making transgenic mice. Cell 1989; 59: 239-41.

[4]   Brinstiel ML, Busslinger M. Dangerous Liaisons: spermatozoa as natural vectors for foreign DNA. Cell 1989; 57: 701-02.

[5]   Chen TM, Chen Y-H. Transgenic sperm or deadly missiles? Fertil Steril 1996; 66: 167.

[6]   Sciamanna I, Vitullo P, Curatolo A, *et al.* Retrotransposons, reverse transcriptase and the genesis of new genetic information. Gene 2009; 448, 180-186

[7]   Webster NL, Forni M, Bacci ML, *et al.* Multi-transgenic pigs expressing three fluorescent proteins produced with high efficiency by sperm mediated gene transfer. Mol Reprod Devel 2005; 72: 68-76

[8]   Vasicek D, Vasickova K, Parkanyi V, *et al.* Effective generation of genetically modified rabbits by sperm mediated gene transfer. World Rabbit Sci 2007; 15: 161-66.

[9]   Zhao YJ, Wei H, Wang Y, *et al.* Production of Transgenic Goats by Sperm-mediated Exogenous DNA Transfer Method. Asian Austral J Anim 2010; 23: 33-40.

[10]   Maione B, Pittoggi C, Achene L, *et al.* Activation of endogenous nucleases in mature sperm cells upon interaction with exogenous DNA. DNA Cell Biol 1997; 16: 1087-97.

[11]   Sciamanna I, Piccoli S, Barberi L, *et al.* DNA dose and sequence dependence in sperm-mediated gene transfer. Mol Reprod Dev 2000; 56: 301-05.

[12]   Zaccagnini G, Maione B, Lorenzini R, *et al.* Increased production of mouse embryos in *in vitro* fertilization by preincubating sperm cells with the nuclease inhibitor aurintricarboxylic acid. Biol Reprod 1998; 59: 1549-53.

[13]   Arezzo F. Sea urchin sperm as a vector for foreign genetic information. Cell Biol Int Rep 1989; 13: 391-04.

[14]   Horan R, Powell R, McQuaid S, *et al.* Association of Foreign DNA with porcine spermatozoa. Arch Androl 1991; 26: 83-82.

[15]   Sperandio S, Lulli V, Bacci ML. Sperm-mediated DNA transfer in bovine and swine species. Anim Biotechnol 1996; 7: 59-77.

[16]   Hogan B, *et al.* Manipulating the mouse embryo (Cold Spring Harbor Laboratory, 1994).

[17]   Zoraqi G, Spadafora C. Integration of foreign DNA sequences into mouse sperm genome. DNA Cell Biol 1997; 16: 291-00.

[18]   Magnano AR, Giordano R, Moscufo N, *et al.* Sperm/DNA interaction: Integration of foreign DNA sequences in the mouse sperm genome. J Reprod Immunol 1998; 41: 187-96.

[19]   Rassoulzadegan M, Leopold P, Vailly J, *et al.* Germ line transmission of autonomous genetic elements in transgenic mouse strains. Cell 1986; 46: 513-19.

[20]   Elbrecht A, Demayo FJ, Tsai MJ, *et al.* Episomal maintenance of a bovine papilloma-virus vector in transgenic mice. Mol Cell Biol 1987; 7: 1276-79.

[21]   Sudo K, Ogata M, Sato Y, *et al.* Cloned origin of DNA-replication in c-myc gene can function and be transmitted in transgenic mice in an episomal state. Nucleic Acids Res 1990; 18: 5425-32.

[22]   Rottmann OJ, Antes R, Hofer P, *et al.* Lipsome mediated gene-transfer *via* spermatozoa into avian egg cells. J Anim Breed Genet 1992; 109: 64-70.

[23]   Khoo HW, Ang LH, Lim HB, *et al.* Sperm cells as vectors for introduction of foreign DNA into zebrafish. Aquaculture 1992; 107: 1-19.

[24]   Bishop JO. Chromosomal insertion of foreign DNA. Reprod Nutr Dev 1996; 36: 607-18.

[25]   Smith K. Theoretical Mechanisms in Targeted and Random Integration of Transgene DNA. Reprod Nutr Dev 2001; 41: 465-85.

[26]   Brinster RL, Braun RE, Lo D, *et al.* Targeted correction of a major histocompatibility class II Eα gene by DNA microinjection into mouse eggs. Proc Natl Acad Sci USA 1989; 87: 3210.

[27]   Gordon JW, Ruddle FH. DNA-mediated genetic-transformation of mouse embryos and bone-marrow - a review. Gene 1985; 33: 121-36.

[28]   Alshawi R, Kinnaird J, Burke J, *et al.* Expression of a foreign gene in a line of transgenic mice is modulated by a chromosomal position effect. Mol Cell Biol 1990; 10: 1192-98.

[29]   Patthy L. Introns and exons. Curr Opin Struct Biol 1994; 4: 383-92.

[30]   Moran JV, Deberardinis RJ, Kazazian HH. Exon shuffling by l1 retrotransposition. Science 1999; 283: 1530-34.

[31]   Capecchi MR. Altering the genome by homologous recombination. Science 1989; 244: 1288-92.

[32]   Reiter LT, Liehr T, Rautenstrauss B, *et al.* Localization of mariner DNA transposons in the human genome by PRINS. Genome Res 1999; 9: 839-43.

[33]   Siyanova EY, Mirkin SM. Expansion of trinucleotide repeats. Mol Biol 2001; 35: 168-82.

[34]   Oliver KR, Greene WK. Transposable elements: powerful facilitators of evolution. BioEssays 2009; 31: 703-14.

[35]   Kaessmann H. Origins, evolution, and phenotypic impact of new genes. Genome Res 2010; 20: 1313-26.

[36]   Smith K. The role of sperm-mediated gene transfer in genome mutation and evolution. Med Hypotheses 2002; 59: 433-37.

CHAPTER 4

Sperm-Mediated Gene Transfer: Implications for Biotechnology and Medicine

Michael Hölker[1,*], Nasser Ghanem[2], Dawit Tesfaye[1] and Karl Schellander[1,*]

[1]University of Bonn, Germany and [2]Cairo University, Egypt

Abstract: Sperm mediated gene transfer (SMGT) was developed as an alternative technique for the production of transgenic animals. This technique is based on the ability of spermatozoa to take up exogenous genes of interest in the form of DNA molecules *in vitro* and deliver them to the oocyte during fertilisation. Thus, novel genetic information could be integrated into the embryo genome in order to alter the expression of specific genes of the offspring and subsequent generations. DNA uptake by spermatozoa is a very specific and well regulated mechanism. Although SMGT has been shown to be efficient, protocols for animal transgenesis are still under optimisation. Recent modifications of SMGT protocols, including intracytoplasmic sperm injection derived transgenesis (ICSI-Tr) and testis mediated gene transfer (TMGT), have been reported. Further understanding of the mechanisms involved in SMGT will enhance our understanding of the biology of fertilisation. Although not yet perfect, the technique of SMGT is of high biotechnological and medical potential. The use of SMGT to generate transgenic domestic animals could enhance their performance, and could also enable the production of proteins and pharmaceuticals within the milk of farm mammals. In addition, it could be used to generate animals as models for human diseases or to produce multitransgenic animals for xenotransplantation purpose. Finally, SMGT also holds promise in the context of human gene therapy in future.

Keywords: Transgenic biotechnology, Gene transfer, Biotechnological research transge ICSI, Xenotransplantation, SMGT, Bioreactors for proteins, ICSI-Tr, Animal desease model, Transgenetic animals, Gene therapy.

INTRODUCTION

Transgenic biotechnology has been well established and widely used to produce genetically modified animals for scientific, pharmaceutical, and agricultural purposes [1-3]. In agriculture its practical applications were predominantly in enhancing livestock production including improved milk production and composition, increased growth rate and disease resistance, improved feed usage and carcass composition, enhanced reproductive performance and increased prolificacy [4, 5]. Today, the most widely used methods for the production of transgenic farm animals are direct microinjection of foreign DNA into the pronuclei of fertilised eggs, nuclear transfer using genetically modified embryonic or somatic donor cells and viral-based constructs as vectors for the introduction of exogenous DNA into embryos. The first transgenic livestock were born more than 2 decades ago [6] and there are numerous potential transgenic methodologies to generate transgenic animals. These methods have been restricted in part by inefficiency when applied in livestock. Moreover, it must be noted that the use of retroviral vectors is affected by safety issues [7]. Thus, there was a need to develop a new methodology that could show high efficiency, being less laborious, at low costs and being safer. In 1971, the first report made by Brackett and his colleagues [8] from Pennsylvania University proved the ability of rabbit spermatozoa to transfer foreign DNA into oocytes. Thereafter, two independent reports confirmed that sperm cells could be carriers for exogenous DNA molecules and those spermatozoa are able to transfer these molecules during fertilisation into oocytes, resulting in genetically modified (transgenic) offspring [9, 10]. The major benefits of sperm mediated gene transfer (SMGT) were found to be high efficiency, low cost and ease of use compared to other methods [10-13]. Another interesting aspect of the use of sperm as DNA vectors is referred to as *en masse*

---

*Address correspondence to Michael Hölker and Karl Schellander: Institute of Animal Science, Animal Breeding and Husbandry Group, University of Bonn, Endenicher Allee 15, 53115 Bonn, Germany; E-mails: ksch@itw.uni-bonn.de; mhoe@itz.uni-bonn.de

transgenesis. In contrast to microinjection, which requires individual manipulation of the embryos, the genetic transformation of a great number of embryos can be obtained collectively by SMGT in one step. This could be of particular interest to transgenesis of aquatic animals including fish [14, 15].

IMPLICATION OF SMGT FOR BIOTECHNOLOGICAL RESEARCH

Although SMGT has been shown to offer efficient, rapid and low-cost protocols for animal transgenesis, it is still under continuous optimisation and modification. Understanding of the mechanism of SMGT in more detail will enhance our understanding of the biological processes occurred at fertilisation. Moreover, since sperm cells can behave as vectors for foreign genetic sequences, sexually reproducing animals may be exposed to alteration by exogenous genetic sequences taken up and carried by sperm cells, with important implications for evolutionary processes [12]. The subsequent fate of sperm-bound DNA, after delivery to the oocyte, is still a contradictory issue [16]. In particular, the question of whether foreign molecules of nucleic acids become integrated into the host genome or remain as extrachromosomal structures is still unsolved [17]. Integration seems to be the favoured outcome when using protocols bypassing direct interaction between the exogenous nucleic acid molecules and the sperm membrane and this is the case when using lipofection [18] or ICSI derived transgenesis (ICSI-Tr, or 'transgenICSI') [19]. Intracytoplasmic sperm injection (ICSI) is a powerful technique in the field of assisted reproduction (ART) and provides exciting opportunities for studying the basic mechanisms of fertilisation and early embryo development. A method involving intracytoplasmic injection of sperm incubated with foreign DNA has been successfully developed for mouse, which has a similar transgenic efficiency to pronuclear microinjection [23]. If large fragments of DNA are needed to be inserted into livestock for correct expression of transgenes, ICSI-Tr might be more successful in generating such animals where other gene insertion techniques prove inadequate [20-22]. During ICSI-Tr mouse spermatozoa are demembranated either by freeze-thawing or by treatment with a detergent such as TritonX-100, then incubated with linear, double stranded DNA that contains the transgene. The rationale for this method was that the exposed perinuclear theca of the sperm head would interact with the DNA and act as a carrier for the transgene. This sperm-DNA complex is then injected into mature metaphase II-oocytes by ICSI, allowing the transgene to be incorporated into the embryonic genome *via* DNA repair mechanisms [24]. The transfection efficiency of this procedure was initially reported as 2.5% (on average) of oocytes injected, or 20% of animals born, with very little mosaicism [23]. Recently, a more efficient version of this method was reported where the efficiencies of oocytes injected and animals born were 4.6% and 45%, respectively [20]. Large DNA inserts (*e.g.* yeast artificial chromosomes) have the advantage that they frequently result in correct gene expression, regardless of position. The transgene expression on the large constructs mimics the endogenous expression pattern of the homologus locus, because their large size usually ensures the inclusion of all regulatory elements [21]. Recently, one transgenic pig was produced through the ICSI-Tr transfer [25].

Other approaches have been developed to create transgenic animals using spermatozoa as vectors. One of these is the testis mediated gene transfer (TMGT) approach which is considered as a simplified version of SMGT, since it does not require IVF or embryo transfer (ET) procedures. *In vivo* studies demonstrated that direct gene delivery into the testis is practicable for gene transfer [26-28]. When a plasmid DNA/liposome complex was injected into the testis of mature male mice which were mated with superovulated females 2-4 days later, it was found that transfection of spermatozoa (epididymal spermatozoa) leads to a high effciency of gene delivery [28] to mid-gestational fetuses (50-100%). A group of researchers led by Chang [29] first examined this mechanism using confocal microscopy of frozen sections of an epididymis prepared 4 days after testis injection with fluorescence-labelled DNA, and demonstrated that the exogenous DNA is bound to the surface of spermatozoa in the cauda epididymis. However, it remains unclear how the introduced DNA reach the epididymal portion or which route is used upon transfer of the DNA from testis to epididymis. An early trial has been performed in order to apply this technique to improve animal performance [29]. In this experiment, a transgene based on the human growth hormone receptor (hGHR) gene labelled with FITC was prepared, the construct was mixed with cationic liposome and injected into rat testis. Exogenous DNA injected with liposomes was bound to the surface of spermatozoa in the cauda epididymis. Afterwards, exogenous DNA introduced into a testis is transferred to epididymis epithelial cells within 4 days and then incorporated to epididymal spermatozoa. A system of germline chimera production that operates *via* the testes rather than through developing embryos which allow fertile spermatozoa to be produced by transfer of testicular cells into juvenile or adult testes has already been developed for chicken transgenesis [30].

IMPLICATIONS OF SMGT FOR BIOTECHNOLOGY

The ability to introduce a specific gene into farm animals, the incorporation of such genes and their stable transmission into the genome of the next generation will enable major genetic advances to be realized in animal breeding. Production of transgenic livestock provides a method to rapidly introduce "new" genes into livestock species without crossbreeding [30-32]. There are numerous potential practical applications of transgenesis in agriculturally important livestock. These applications include improving production and composition, increasing growth rate, improving feed utilisation, improving carcass composition, increasing disease resistance, enhancing reproductive performance, increasing prolificacy, as well as engineering cell and tissue characteristics for biomedical research [32].

Improvement of Livestock Traits

As demands for protein continue to increase through expanding markets and increased population growth, additional gains in protein production efficiency has to occur if the animal industry is to maintain a substantial portion of the world market. Because increases in protein deposition become more difficult to achieve through traditional means, animal scientists could develop new techniques to alter animal growth and muscle composition based on understanding of the biological mechanisms that control the growth of muscle. In addition, more attention should be given toward the quality of the animal products as this is critical to consumer acceptance with increasing human welfare. The emergence of the first functional transgenesis in laboratory mice was first reported in 1982, when Palmiter and his colleagues [33] produced transgenic mice expressing rat growth hormone fused to the metallothionein promoter sequence, with the animals displaying increased growth over the normal rate. This report opened the possibility of using transgenesis as an instrument to increase meat production. In swine several attempts have been made to improve growth and meat composition by the introduction of transgenes. In one study, expression of an exogenous insulin-like growth factor gene in the muscle of pigs resulted in a significant reduction in fat and in an increase in lean muscle in gilts but not in boars [34]. In another study, a widely expressed exogenous growth hormone gene tended to increase live weight gain, improved feed efficiency and reduced back fat thickness [35]. However, transgenic pigs expressing the growth hormone exhibited only a slight increase in growth and a high incidence of collateral effects, such as gastric ulcers, arthritis, cardiomegaly, dermatitis, and renal disease [36]. The use of transgenic technologies to modify feed efficiency was also performed by introduction of a phytase gene in pigs in an attempt to increase the bioavailability of phosphorus from phytic acid in corn and soybean products. The production of transgenic pigs expressing salivary phytase as early as 7 days of age has been reported [37]. The salivary phytase provided complete digestion of the dietary phytate phosphorus resulting in a reduction of phosphorus output by up to 75%. Furthermore, these transgenic pigs required almost no inorganic phosphorus supplementation to the diet to achieve normal growth. The use of phytase transgenic pigs in commercial pork production could result in decreased environmental phosphorus pollution from livestock operations.

Milk protein, 80% of which consists of casein, is one of the most valuable components of milk because of its nutritional value and processing properties. Therefore, casein is a prime target to improve milk composition. Additional copies of bovine beta and kappa casein genes have been inserted into bovine female fibroblasts followed by somatic cell nuclear transfer using four independent donor cell lines to generate transgenic calves [38]. The transgenic offspring were evaluated for milk production and composition and showed substantial expression and secretion of the transgene-derived caseins into the milk. Additionally, transgenic offspring had an 8 to 20% increase in beta casein and a two-fold increase in kappa casein.

Mastitis is well known as the most costly disease in animal agriculture, resulting in decreased profits all over the world annually. Economic losses associated with mastitis are both direct (loss in milk production, veterinarian's time, herdsman's time, cost of drugs and cost of discarded milk) and indirect (high culling rate, extended calving intervals and reduced milk quality). Therefore it is of great importance to protect animals from this disease. On thinkable approach is to produce transgenic cows that are able to synthesise substances that give more resistance against mastitis. Transgenic cows secreting lysostaphin, a natural substance that is able to inhibit the activity of *Staphylococcus aureus* (*S. aureus*) which is one of the main have been produced [39]. The majority of transgenic females expressing lysostaphin in their milk were resistant to intramammary infections when challenged with *S. aureus*.

Pathogenic infections have a significant negative impact on neonatal survival of newborn piglets. Therefore, piglets fed with human lysozyme, a natural antibiotic, might be more resistant to gastrointestinal infections. Transgenic swine expressing recombinant human lysozyme were generated by somatic cell nuclear transfer [40]. Three cloned female pigs were born and one of them expressed human lysozyme 50-fold higher compared to the pigs' native lysozyme. Therefore, introducing human lysozyme into pigs' milk could bear a potential to benefit the piglets by defending against pathogenic bacteria resulting in increased newborn survival rates.

The ability to express transgenes in milk-producing animals has resulted in the creation of "bioreactors" - animals that produce large amounts of a given recombinant protein in their milk, in fully biologically active form through proper posttranslational modification (PTM), for purification and therapeutic use. Several such animals have been engineered to date [41]. Most of these animals have been produced by standard methods of transgenesis, although some animals have been produced using forms of SMGT. Transgenic rabbits, for example, were produced by direct injection of a DMSO–DNA complex *in vivo* into the testes as an attempt to apply SMGT in mammary gland bioreactor [42]. Interestingly, 17 out of 21 of transgenic female rabbits (81%) expressed human lactoferrin (LF) protein in their glands as detected *via* southern blotting.

Bioreactors for Proteins

The production of valuable pharmaceutical human enzymes, hormones, antibodies and growth factors currently requires large-scale cell cultures to generate products in biological systems. This approach requires the production of transfected eukaryotic or bacterial cell lines which contain transgene constructs for the generation of recombinant gene products. However, the necessary posttranslation processing and propper folding does not occur in these organisms, thus frequently rendering many mammalian proteins non-functional. Moreover, the production process for recombinant proteins in mammalian cell culture *in vitro* is very expensive due to the requirement of pathogen-free conditions with constant monitoring, buffering and temperature regulation in the medium. Therefore, the concept of "pharmaceutical farming", or "pharming", where large transgenic animals are used as bioreactors for protein production, is very appealing to the pharmaceutical industry. Mammary gland transgene expression currently is the preferred option because it allows mass production of large amounts correctly processed proteins in a temperature-regulated fluid that may be collected daily in a non-invasive fashion [43]. Through genetic engineering it has become possible to produce any protein from any animal, plant or bacterial species in the milk of mammals [44]. For example, it is possible to express milk proteins and other proteins of pharmaceutical value in the milk of mice, rabbits, pigs, goats and sheep [45-49].

SMGT and ICSI-Tr offer an opportunity to insert large transgenes into livestock animals for the correct expression and processing of gene products. Producing large quantities of biological products in animal bioreactors may help to alleviate the big demand for biomolecules that are currently synthesized by very expensive procedures. For example, Hemophilia A is an inherited, sex-linked bleeding disease resulting from a defective or deficient coagulation factor VIII (FVIII). With an incidence ratio of 1:5000 in male births, hemophilia A comprises the majority of hemophilia patients (approximately 80%). Recently, sperm mediated gene transfer (SMGT) techniques has been used to generate transgenic mice capable of producing FVIII protein, in order to produce clotting factor VIII concentrate for human use [50]. The efficiency rate was 33.3% (3/9), which was consistent with the efficiency of SMGT reported in previous publications (37.5% in pigs, 33% in mice and 57.1% in rabbits, respectively) [42, 51].

IMPLICATIONS OF SMGT FOR MEDICAL PURPOSE

Gene-based biomedical research offers one of the best hopes yet for curing the major diseases which still afflict mankind. The use of transgenic animals which produce human drugs will become more common. Global demand continues to grow for human proteins and vaccines, which serve numerous therapeutic purposes such as treatments for cystic fibrosis, hemophilia, osteoporosis, arthritis, malaria, and human immunodeficiency virus. Transgenic animals can also produce monoclonal antibodies (antibodies specifically targeted toward disease proteins) used in the development of vaccines. In addition, transgenesis allows the generation of animal models for human diseases. Finally, human gene therapy (HGT), the treatment or prevention of disease by gene transfer is, regarded by many as a potential revolution in medicine, because gene therapies target the causes of disease, whereas most current drugs treat the symptoms. In all the foregoing applications, SMGT holds significant promise.

Transgenic Animals as a Model for Human Diseases

The well-characterized physiology, genetics and short lifespan of mice allow for rapid analysis of the phenotypic changes associated with the transgene over their entire lifespan. These characteristics facilitate the accelerated development of new diagnostic and therapeutic treatments for human diseases. Therefore, many scientists have generated mutant mice defective in the expression of one or more genes through a variety of methods which have most commonly utilized random mutagenesis followed by phenotypic and then genotypic analysis. There are many transgenic rodents that model human diseases such as sick cell anaemia [52], AIDS [53], amylotropic lateral sclerosis spondylitis [54] and cancer [55]. One strategy is to disrupt an endogenous gene by random insertion of a transgene into the host's genome, the consequences of which can then be studied. This action can be mediated by several methods, including SMGT. Alternatively, 'knockout' and 'knockin' animals can be produced using targeted homologous recombination in ES cells or somatic cells for nuclear transfer. However, this approach is generally expensive, complex, and time consuming [56]. RNA interference may provide a more viable alternative because it is relatively simple, inexpensive, faster, and may in fact provide a more suitable model in those diseases where decreased expression of the gene (as opposed to a zero expressing mutation) accounts for certain disease phenotypes [56]. Moreover, there is no need for a locus specific integration of transgenes coding for RNAi sequences. If one makes an interfering or antisense construct to an inhibitory protein such as growth differentiation factor-8 (GDF-8, also known as myostatin) one could potentially increase skeletal muscle growth [57]. Moreover, an increased level of insulin-like growth factor-I (IGF-I) in serum of hGH transgenic mice is another expample of this type of situation [58]. Therefore SMGT could be a viable method to introduce such RNAi transgenes and overall to produce genetic disease models due to decreased or enhanced expression of certain genes.

Transgenic Animals for Xenotransplantation

The ability to create transgenic large animals by SMGT with high yields of positive founders at relatively low cost compared with microinjection is important when one needs to create multitransgene animals. SMGT will also be of benefit to those whose work requires the use of transgenic animal models in medicine [59]. Swine have been used in biomedical applications for many decades as a model for human disease processes, as a genetically defined model for surgery and transplantation, and as a source of human therapeutics. Thousands of patients die every year for lack of a replacement heart, liver, or kidney. Clinical use of transplantation has become one of the major treatments for many diseases associated with terminal organ failure. However, the success rate is limited by lack of human organ supply, which has greatly limited the number of patients who can receive such life-saving treatment. The pig is the most likely donor animal for xenotransplantation of organs, but may well require multiple transgenes to be a satisfactory donor for humans [60, 61]. Given the high efficiency of transgenesis, SMGT could greatly facilitate the production of such pigs [59]. In 2002 Lavitrano and his colleagues described the production of transgenic pigs that express human decay accelerating factor (hDAF) using sperm-mediated gene transfer. This factor plays an important role in overcoming the first rejection barrier to pig-to-primate transplants, so that this group of researchers elected to generate pigs expressing hDAF by SMGT as their starting point for studies on xenotransplantation. The several *hDAF* transgenic founders and first generation expressed *hDAF* at the mRNA and protein levels in all tissues examined. The expressed protein confers resistance against the action of human complement. The efficiency of creating transgenic pigs by SMGT is reportedly significantly greater than by using microinjection. In eight experiments, 53 of 93 pigs generated were transgenic (57%). This result contrasts with reported efficiencies of 0.5–4% in pigs using microinjection [62], *i.e.*, a 25-fold improvement in efficiency. The *hDAF* gene detected was human specific *DAF*, and not swine *DAF*, this was based on the high stringency used with an *hDAF* probe, on the absence of any sequence detected in nontransgenic pigs, and was supported by the observation that fragments of different sizes were present in the transgenic founder animals on Southern blots [59]. The key feature of this study was that, by this method, a large number of pigs have been obtained which had the *hDAF* minigene integrated into the genome and which correctly expressd the hDAF protein on the cell membranes of the desired tissues. The efficiency of transgenesis obtained by SMGT is greatly superior to that obtained by any other methods [10, 63] which holds promise for studies involving large animal models.

To produce multitransgenic animals for successful xenotransplantation a simple and efficient method is required. A study was done to investigate whether multiple transgenes could be introduced by SMGT into swine using the expression of three fluorescent proteins; enhanced blue (EBFP), green (EGFP), and red (DsRed2) fluorescent proteins [13]. High frequency of DNA incorporation using the three colour genes was observed. About 171 of 195 embryos (88%) normally developed to the morula/blastocysts stage and were examined at day 6 post inseminations. Virtually all expressed all three fluorescent proteins. Genomic DNA of 18 piglets born from two litters was screened by PCR, showing that all piglets were transgenic with at least one gene. Of these 18 piglets, 7 were triple transgenic, another 7 were double transgenic, and 4 piglets were single transgenic. Recently, a study was performed in pigs to understand the molecular mechanisms of acute rejection in xenotransplantation [64]. In this work, three integrative constructs carrying three constructs were incubated with porcine sperms to perform SMGT.

Each construct contained a human gene that is considered to be a key modulator of inflammation, namely hemeoxygenase 1 (hHO1), ectonucleoside triphosphate diphosphohydrolase 1 (hENTPD1, also known as hCD39), and 5′-ecto-nucleotidase (hNT5E, also known as hCD73). After *in vitro* fertilisation and 7 days of embryo culture, embryos at morula/blastocyst stage were evaluated for correct embryo development. Among 1048 matured oocytes fertilised with sperm cells that had been incubated with these 3 integrative constructs, a total of 775 were subsequently considered presumptive zygotes among which 340 embryos (48%) developed into blastocysts. Among the 150 transgenic blastocysts (73%), 72 were single (48%), 47 double (31%), and 31 (21%) triple transgenic. The authors concluded that such an *in vitro* system could be a useful tool to screen transgenic constructs before their use for the production of transgenic animal models.

Gene Therapy

The most obvious potential application of SMGT in the context of human gene therapy would be as a means to genetically modify the human germline – *i.e.* to permit human germline genetic modification (HGGM). However, SMGT methodology is clearly at much too early a stage of development to permit its use for this purpose at present. Moreover, the fact that it is not existing patients or people who would be affected, but rather their offspring – and subsequent generations – renders HGGM ethically problematic.

By contrast, somatic gene therapy, *i.e.* the treatment or prevention of diseases in existing patients using gene transfer, has already been attempted (with some successes) in a number of clinical trials, and is regarded by many as being far more ethically acceptable than HGGM. Somatic gene therapy may be viewed as a potential revolution in medicine, because gene therapies target the causes of disease, whereas most current drugs treat the symptoms. Transgene delivery to the various organs of experimental model animals has become a mainstay of preclinical gene therapy research, and *in vivo* gene transfer techniques have become popular tools in gene therapy attempts. SMGT, although not applicable to many conditions (*i.e.* those not concerning the male gonad), may be able to offer a role as a gene therapy approach towards treating some forms of male infertility.

Genetic analyses in infertile male patients have shown that gene domain azoospermia factor (AZF) is involved in spermatogenesis [65, 66]. Deleted or non-functional AZF genes, and possibly other genes on the human Y chromosome, are a cause of azoospermia [67, 68]. Therefore, idiopathic male infertility might be associated with several gene deficiencies and, as for other diseases, this might be treatable by gene therapy in the future [69, 70]. Gene transfer to mouse testes (TMGT) using electroporation and its influence on spermatogenesis has been studied in this context [71]. Results of this study demonstrated that damage to spermatogenesis from testis gene transfer by electroporation is only temporary. The fertility of gene transfer treated male mice was also evaluated by mating them with normal female mice followed by counting the number of offspring. The number of offspring did not differ significantly for males before and after gene transfer and all transfected males fathered normal offspring. These results suggests that TMGT by electroporation might be effective for transfecting testicular cells and could be applicable for *in vivo* gene therapy for male infertility in the future. Viral-vector mediated testis gene transfer can be used for correcting sertoli cell dysfunction in mice [72]. In this study, *Sl/Sld* mutant male mice which were infertile due to sertoli cell dysfunction were used. This research group has reported that not only was spermatogenesis restored in all recipient testes, but also spermatozoa collected from transduced testes were able to generate normal pups after microinsemination.

Impaired spermatogenesis is an adverse effect of cryptorchidism in many mammals. Erythropoietin (Epo), a hematopoietic cytokine, regulates erythrocyte production by acting on the proliferation, differentiation, and apoptosis of erythroid progenitor cells [73]. This protein is one of the promising bio-therapeutics that could be used to overcome the problem of impaired spermatogenesis due to cryptorchidism due to its ability to stimulate steroidogenesis in the testis and increased testosterone production *in vitro* [74] and *in vivo* [75] of experimental animals. Therefore, a study was performed investigating the effects of rat Epo on spermatogenesis by transferring rat Epo genes into cryptorchid testes by means of *in vivo* electroporation. The transfer of erythropoietin gene constructs into the testes of cryptorchid male rats by *in vivo* electroporation showed the reversal of germ cell loss associated with cryptorchidism [76].

CONCLUDING REMARKS

SMGT has been demonstrated as a workable method for the production of transgenic animals. Although protocols are still under optimisation, the technique is potentially of high biotechnological and medical value. SMGT can produce transgenic animals to improve their performance, to generate animals as human disease models and to facilitate the production of pharmaceuticals. Moreover the technique holds future promise in the context of human gene therapy.

REFERENCES

[1]   Mullins LJ, Mullins JJ. Transgenesis in the rat and larger mammals. J Clin Invest 1996; 97: 1557-60.
[2]   Chan AWS. Transgenic animals: Current and alternative strategies. Cloning 1999; 1: 25-46.
[3]   Wall RJ. New gene transfer methods. Theriogenology 2002; 57: 189-01.
[4]   Houdebine L-M. Transgenesis to improve animal production. Livestock Prod Sci 2002; 74: 255-68.
[5]   Bacci ML. A brief overview of transgenic farm animals. Vet Res Commun 2007; 31: 9-14.
[6]   Hammer RE, Pursel VG, Rexroad CE, Jr, *et al.* Production of transgenic rabbits, sheep and pigs by microinjection. Nature 1985; 315: 680-83.
[7]   Lavitrano M, Busnelli M, Cerrito MG, *et al.* Sperm-mediated gene transfer. Reprod Fertil Dev 2006; 18: 19-23.
[8]   Brackett BG, Baranska W, Sawicki W, *et al.* Uptake of heterologous genome by mammalian spermatozoa and its transfer to ova through fertilisation. Proc Natl Acad Sci USA 1971; 68: 353-57.
[9]   Arezzo F. Sea-urchin sperm as a vector of foreign genetic information. Cell Biol Int Rep 1989; 13: 391-404.
[10]   Lavitrano M, Camaioni A, Fazio VM, *et al.* Sperm cells as vectors for introducing foreign DNA into eggs: genetic transformation of mice. Cell 1989; 57: 717-23.
[11]   Lavitrano M, French D, Zani M, *et al.* The interaction between exogenous DNA and sperm cells. Mol Reprod Dev 1992; 31: 161-69.
[12]   Smith K, Spadafora C. Sperm-mediated gene transfer: applications and implications. Bioessays 2005; 27: 551-62.
[13]   Webster NL, Forni M, Bacci ML, *et al.* Multi-transgenic pigs expressing three fluorescent proteins produced with high efficiency by sperm mediated gene transfer. Mol Reprod Dev 2005; 72: 68-76.

[14] Collares T, Campos VF, Seixas FK, *et al.* Transgene transmission in South American catfish (Rhamdia quelen) larvae by sperm-mediated gene transfer. J Biosci 2010; 35: 39-47.

[15] Spadafora C. Sperm cells and foreign DNA: a controversial relation. Bioessays. 1998; 20: 955-64.

[16] Hoelker M, Mekchay S, Schneider H, *et al.* Quantification of DNA binding, uptake, transmission and expression in bovine sperm mediated gene transfer by RT-PCR: effect of transfection reagent and DNA architecture. Theriogenology 2007; 67: 1097-107.

[17] Spadafora C. Sperm-mediated 'reverse' gene transfer: a role of reverse transcriptase in the generation of new genetic information. Hum Reprod 2008; 23: 735-40.

[18] Harel-Markowitz E, Gurevich M, Shore LS, *et al.* Use of sperm plasmid DNA lipofection combined with REMI (restriction enzyme-mediated insertion) for production of transgenic chickens expressing eGFP (enhanced green fluorescent protein) or human follicle-stimulating hormone. Biol Reprod 2009; 80: 1046-52.

[19] Naruse K, Ishikawa H, Kawano H, *et al.* Production of transgenic pig expressing human albumin and enhanced green fluorescent protein. J Reprod Dev 2005; 51: 539-46.

[20] Moreira PN, Giraldo P, Cozar P, *et al.* Efficient generation of transgenic mice with intact yeast artificial chromosomes by intracytoplasmic sperm injection. Biol Reprod 2004; 71: 1943-47.

[21] Moreira PN, Pozueta J, Perez-Crespo M, *et al.* Improving the generation of genomic-type transgenic mice by ICSI. Transgenic Res 2007; 16: 163-68.

[22] Osada T, Toyoda A, Moisyadi S, *et al.* Production of inbred and hybrid transgenic mice carrying large (>200kb) foreign DNA fragments by intracytoplasmatic sperm injection. Mol Reprod Dev 2005; 72: 329-35.

[23] Perry ACF, Wakayama T, Kishikawa H, *et al.* Mammalian transgenesis by intracytoplasmic sperm injection. Science 1999; 284: 1180-83.

[24] Perry ACF. Hijacking oocyte DNA repair machinery in transgenesis? Mol Reprod Dev 2000; 56: 319-24.

[25] Kurome M, Ueda H, Tomii R, *et al.* Production of transgenic-clone pigs by the combination of ICSI-mediated gene transfer with somatic cell nuclear transfer. Transgenic Res 2006; 15: 229-40.

[26] Blanchard KT, Boekelheide K. Adenovirus-mediated gene transfer to rat testis *in vivo*. Biol Reprod 1997; 56: 495-00.

[27] Yamazaki Y, Fujimoto H, Ando H, *et al. In vivo* gene transfer to mouse spermatogenic cells by deoxyribonucleic acid injection into seminiferous tubules and subsequent electroporation. Biol Reprod 1998; 59: 1439-44.

[28] Sato M, Gotoh K, Kimura M. Sperm-mediated gene transfer by direct injection of foreign DNA into mouse testis. Transgenics 1999; 2: 357-69.

[29] Chang KT, Ikeda A, Hayashi K, *et al.* Possible mechanisms for the testis-mediated gene transfer as a new method for producing transgenic animals. J Reprod Dev 1999; 45: 37-42.

[30] Lee YM, Jung JG, Kim JN, *et al.* A testis-mediated germline chimera production based on transfer of chicken testicular cells directly into heterologous testes. Biol Reprod 2006; 75: 380-86.

[31] Pursel VG, Rexroad CE. Status of research with transgenic farm animals. J Anim Sci 1993; 71: 10-19.

[32] Wheeler MB. Production of transgenic livestock: Promise fulfilled. J Anim Sci 2003; 81: 32-37.

[33] Palmiter RD, Bringster RL, Hammer RE, *et al.* Dramatic growth of mice that develop from egg microinjected with metallothionein-growth hormone fusion genes. Nature 1982;.300: 611-15.

[34] Pursel VG, Mitchell AD, Bee G, *et al.* Growth and tissue accretion rates of swine expressing an insulin-like growth factor I transgene. Anim Biotechnol 2004; 15: 33-45.

[35] Nottle MB, Nagashima H, Verma PJ, *et al.* Production and analysis of transgenic pigs containing a metallothionein procine growth hormone gene construct. In: Transgenic Animals in Agriculture. CAB International: Wallingford. NY, 1999; p. 145.

[36] Pursel VG, Pinkert CA, Miller KF, *et al.* Genetic engineering of livestock. Science 1989; 244: 1281-88.

[37] Golovan SP, Meidinger RG, Ajakaiye A, *et al.* Pigs expressing salivary phytase produce low-phosphorus manure. Nat Biotechnol 2001; 19: 741-45.

[38] Brophy B, Smolenski G, Wheeler T, *et al.* Cloned transgenic cattle produce milk with higher levels of beta-casein and kappa-casein. Nat Biotechnol 2003; 21: 157-62.

[39] Wall RJ, Powell AM, Paape MJ, *et al.* Genetically enhanced cows resist intramammary Staphylococcus aureus infection. Nat Biotechnol 2005; 23: 445-51.

[40] Tong J, Wei H, Liu X, *et al.* Production of recombinant human lysozyme in the milk of transgenic pigs. Transgenic Res 2010; 12 (In press).

[41] Yang X, Carter MG. Transgenic animal bioreactors : a new line of defense against chemical weapons?. Proc Natl Acad Sci USA 2007; 104: 13859-60.

[42]   Li L, Shen W, Min L, *et al*. Human lactoferrin transgenic rabbits produced efficiently using dimethylsulfoxide-sperm-mediated gene transfer. Reprod Fertil Dev 2006; 18: 689-95.

[43]   Kolb AF, Coates CJ, Kaminski JM, *et al*. Site-directed genome modification: nucleic acid and protein modules for targeted integration and gene correction. Trends Biotechnol 2005; 23: 399-406.

[44]   Bremel RD, Yom HC, Bleck GT. Alteration of milk composition using molecular genetics. J Dairy Sci 1989; 72: 2826-33.

[45]   Simons JP, Mc Clenaghan M, Clark AJ. Alteration of the quality of milk by expression of sheep beta-lactoglobulin in transgenic mice. Nature 1987; 328: 530-32.

[46]   Buehler TA, Bruyère T, Went DF, *et al*. Rabbit ß-casein promoter directs secretion of human interleukin-2 into the milk of transgenic rabbits. Biotechnology 1990; 8: 140-43.

[47]   Ebert KM, Selgrath JP, Ditullio P, *et al*. Transgenic production of a variant of human tissue-type plasminogen activator in goat milk: generation of transgenic goats and analysis of expression. Biotechnology 1991; 9: 835-38.

[48]   Wall RJ, Pursel VG, Shamay A, *et al*. High level synthesis of a hererologous milk protein in the mammary glands of transgenic swine. Proc Natl. Acad. Sci. USA 1991; 88: 1696-00.

[49]   Wright G, Garver A, Cottom D, *et al*. High level of expression of active alpha-1-antitrypsin in the milk of transgenic sheep. Biotechnology 1991; 9: 830-34.

[50]   Yin J, Zhang JJ, Shi GG, *et al*. Sperm mediated human coagulation factor VIII gene transfer and expression in transgenic mice. Swiss Med Wkly 2009; 139: 364-72.

[51]   Chang K, Qian J, Jiang MS, *et al*. Effective generation of transgenic pigs and mice by linker based sperm-mediated gene transfer. BMC Biotechnol 2002; 2: 5-17.

[52]   Ryan RM, Townes TM, Reilly MP, *et al*. Human sickle haemoglobin in transgenic mice. Science 1990; 247: 566-68.

[53]   Vogel J, Hinrichs SH, Reynolds RK, *et al*. The HIV tat gene includes dermal lesions resembling Kaposi's sarcoma in transgenic mice. Nature 335: 606-11.

[54]   Gordon JW. Trangenic technology and laboratory animal science. ILAR J 1997; 38: 32-41.

[55]   Sinn E, Muller W, Pattengale P, *et al*. Coexpression of MMTV/v-Ha-ras and MMTV/c-myc genes in transgenic mice: synergistic action of oncogenes *in vivo*. Cell 1987; 49: 465-75.

[56]   Xia XG, Zhou H, Xu Z. Transgenic RNAi: Accelerating and expanding reverse genetics in mammals. Transgenic Res 2006; 15: 271-75.

[57]   Mc Pherron AC, Lawler AM, Le SJ. Regulation of skeletal muscle mass in mice by a new TGF-beta superfamily member. Nature 1997; 387: 83-90.

[58]   Palmiter RD, Norstedt G, Gelinas RE, Hammer RE, Bringster RL. Metallothionein-human GH fusion genes stimulate growth of mice. Science 1983; 222: 809-14.

[59]   Lavitrano M, Bacci ML, Forni M, *et al*. Efficient production by sperm-mediated gene transfer of human decay accelerating factor (hDAF) transgenic pigs for xenotransplantation. Proc Natl Acad Sci USA 2002; 99: 14230-35.

[60]   Cozzi E, White DJ. The generation of transgenic pigs as potential organ donors for humans. Nat Med 1995; 1: 964-66.

[61]   Robson SC, Schulte AM, Esch J, Bach FH. Factors in xenograft rejection. Ann N Y Acad Sci 1999; 875:261-76.

[62]   Niemann H, Kues WA. Transgenic livestock: premises and promises. Anim Reprod Sci 2000; 61: 277-93.

[63]   Lavitrano M, Maione B, Forte E, *et al*. The interaction of sperm cells with exogenous DNA: a role of CD4 and major histocompatibility complex class II molecules. Exp Cell Res 1997; 233: 56-62.

[64]   Vargiolu A, Manzini S, de Cecco M, *et al*. *In vitro* production of multigene transgenic blastocysts *via* sperm-mediated gene transfer allows rapid screening of constructs to be used in xenotransplantation experiments. Transplant Proc 2010; 42: 2142-45.

[65]   Nagafuchi S, Namiki M, Nakahori Y, *et al*. A minute deletion of the Y chromosome in men with azoospermia. J Urol 1993; 150: 1155-57.

[66]   Vogt PH, Edelmann A, Hirschmann P, Köhler MR The azoospermia factor (AZF) of the human Y chromosome in Yq11: function and analysis in spermatogenesis. Reprod Fertil Dev 1995; 7: 685-93.

[67]   Reijo R, Alagappan RK, Patrizio P, Page DC. Severe oligozoospermia resulting from deletions of azoospermia factor gene on Y chromosome. Lancet 1996; 347: 1290-93.

[68]   Ruggiu M, Speed R, Tagarrt M, *et al*. The mouse DAZLA gene encodes a cytoplasmic protein essential for gametogenesis. Nature 1997; 389: 73-77.

[69]   Reed CC, Gauldie J, Iozzo RV. Suppression of tumorigenicity by adenovirus-mediated gene transfer of decorin. Oncogene 2002; 21: 3688-95.

[70]   Voeks D, Martiniello-Wilks R, Madden V, *et al.* Gene therapy for prostate cancer delivered by ovine adenovirus and mediated by purine nucleoside phosphorylase and fludarabine in mouse models. Gene Ther 2002; 9: 759-68.

[71]   Umemoto Y, Sasaki S, Kojima Y, *et al.* Gene transfer to mouse testes by electroporation and its influence on spermatogenesis. J Androl 2005; 26: 264-71.

[72]   Ikawa M, Tergaonkar V, Ogura A, *et al.* Restoration of spermatogenesis by lentiviral gene transfer: offspring from infertile mice. Proc Natl Acad Sci USA 2002; 99: 7524-29.

[73]   Krantz SB. Erythropoietin. Blood 1991; 77: 419-34.

[74]   Foresta C, Mioni R, Bordon P, *et al.* Erythropoietin and testicular steroidogenesis: the role of second messengers. Eur J Endocrinol 1995; 132: 103-08.

[75]   Foresta C, Mioni R, Bordon P, *et al.* Erythropoietin stimulates testosterone production in man. J Clin Endocrinol Metab 1994; 78: 753-56.

[76]   Dobashi M, Goda K, Maruyama H, Fujisawa M. Erythropoietin gene transfer into rat testes by *in vivo* electroporation may reduce the risk of germ cell loss caused by cryptorchidism. Asian J Androl 2005; 7: 369-73.

CHAPTER 5

Methodology of Sperm-Mediated Gene Transfer

Yidong Niu[*]

Peking University People's Hospital, China

Abstract: The capture of exogenous DNA and its transfer to eggs by sperm cells are the two key processes in SMGT. During the past two decades, the efficiency of the capture of exogenous DNA molecules by sperm cells has been increased by various approaches, such as use of liposomes, electroporation, Triton-X or DMSO-treated spermatozoa, viral vectors, and magnetic nanoparticles. Testis-mediated gene transfer (TMGT) is a novel alternative variant of SMGT used to produce transgenic animals *in vivo*. This chapter presents the background and features of both non-viral and viral forms of SMGT, together with TMGT, as developed in recent years. Additionally, modes of fertilisation by transgene-bearing sperm are discussed.

Keywords: Male germ cell, SMGT, TMGT, Viral vector, Liposome, Linker, Electroporation, Magnatofection, IVF, ICSI, AI, GIFT, Nature mating.

INTRODUCTION

The primary technique in sperm-mediated gene transfer (SMGT) can be traced back to 1971, when the ability of capturing foreign DNA molecules by spermatozoa was first reported [1]. In that protocol, the ejaculated rabbit spermatozoa were incubated directly with [^{3}H] thymidine-labeled SV40 DNA, and then the heterologous genomes were identified from sperm cells and ovum after artificial insemination (AI). However, this study has been ignored for a long time until two independent studies rediscovered that spermatozoa could take up foreign DNA molecules and transmit them into oocytes and offspring in the late 1980s [2, 3]. Since 1989, although there were serious doubts and controversy [4, 5], some unusual molecular mechanisms contributing to SMGT have been described, such as capturing of exogenous DNA [6] and RNA [7] molecules by spermatozoa and insertion of foreign fragments into sperm chromosomes [8]. Furthermore, the efficiency of SMGT in some experiments is reported to be much higher than those by other methods and rates as high as 88% [9]. Therefore, SMGT offers great advantage for transgenesis.

The technique of SMGT is simple in principle, involving interactions of exogenous DNA with spermatozoa, but the mechanisms behind this apparently simple process have been demonstrated to be complicated [10]. Since the SMGT method using naked DNA is thought to be not so efficient for transgenesis, other approaches were developed in succession, during the last twenty years, to improve the efficiency of capturing exogenous DNA by spermatozoa, such as lipofection [11], electroporation [12], use of Triton-X or DMSO-treated spermatozoa [13, 14] and application of virus or magnetic nanoparticle vector [15-17]. A novel alternative mean of SMGT is testis mediated gene transfer (TMGT). This *in vivo* technique introduces exogenous DNA molecules directly into testis by injection [18], and rates as high as 41% and 37% of transgenesis can be obtained in F1 and F2 mice offspring [19]. To date, SMGT and/or TMGT have been employed to generate transgenic animals in various species, such as fish [20], amphibian [21], aves [22], mammals [3, 19, 23], and some invertebrates [2, 24, 25].

METHODS FOR UPTAKE OF EXOGENOUS DNA BY MALE GERM CELLS

Current SMGT strategies employ two classes of methods, viral-based and nonviral methods, to promote uptake of exogenous genes by spermatozoa. This chapter presents the backgrounds and features of both the two methods developed for transfection of spermatozoa to produce transgenic mammals by SMGT.

*Address correspondence to Yidong Niu:** Laboratory Animal Unit, Peking University People's Hospital, No.11 South Xizhimen Avenue, Beijing 100044, China; E-mail: niuyd07@yahoo.com.cn

The Viral-Based Methods

Most viral vectors employed in SMGT are derived from adenovirus, retrovirus and lentivirus, although other viral vectors may be used. The basic processes by which viral vectors are introduced to the target cells includes incubation, injection, and other steps, which will be described under nonviral methods.

Adenoviral (Ad) Vectors

Due to the very efficient nuclear entry mechanism of adenovirus (Ad) and its low pathogenicity for hosts, adenoviral vectors have become a popular tool for gene transfer into mammalian cells. Production of Ad vector requires no specific laboratory skills and high titer is easily obtained. When constructed with specific deletions rendering the virus replication deficient, Ad vectors have a relatively large cloning capacity. Unlike other vectors, adenovirus can infect both dividing and nondividing cells in a cell cycle-independent manner. The most commonly used Ad vector is the E1 (early gene 1) deleted (so-called the first generation, Fig. **1A**) Ad vector [26]. The exogenous genes are designed to insert in place of the E1 region, which locates at the left end of the Ad genome. Because E1 products are essential for virus replication, E1-complementing cell lines, such as 911 [27] or PER.C6 [28], must be used to propagate the vectors (Fig. **1B**). If the unnecessary region for replication *in vitro*, E3, is removed, the cloning capacity can be increased to 8.2 kb (Fig. **1A**) [29].

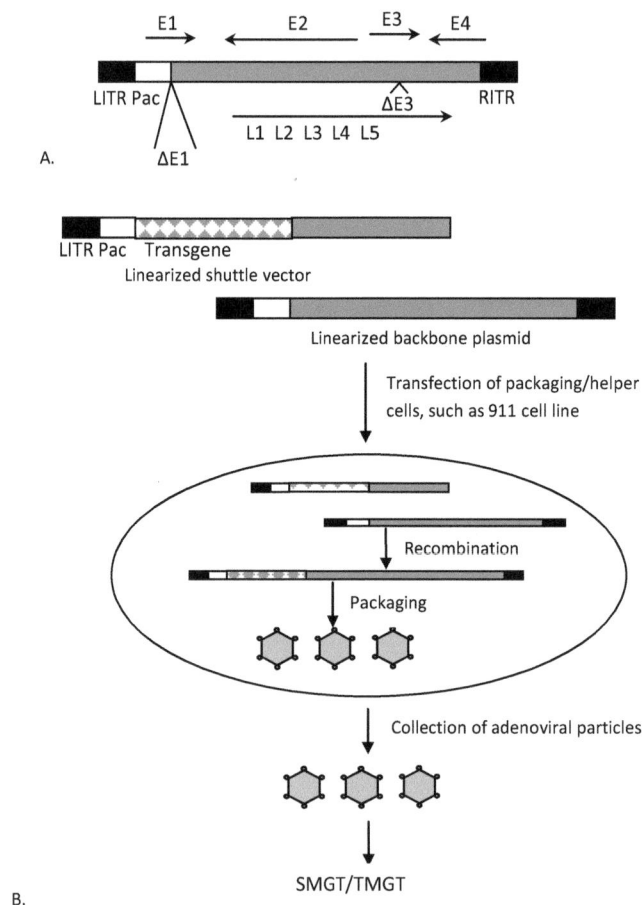

Figure 1: Schematic Structure of First Generation Adenoviral (Ad) Vectors (**A**) and Procedure of Ad Vector Mediated Gene Transfer in SMGT or TMGT (**B**). A. The left and right inverted terminal repeats (ITR) are symbolised by black rectangles. The locations of the early and late transcription units are shown up and under the arrows. The insert sises depends on the sises of deletions indicated by ΔE1 and ΔE3. The maximal transgene capacity of the first generation Ad vectors is about 8.2 kb. B. The shuttle vector and backbone plasmid are linearised and then transfected into helper cell line, such as 911 cell line. Inside the cell, the plasmids recombine and the recombined genome is replicated and packaged into adenoviral particles.

Blanchard and Boekelheide firstly reported successful adenovirus-mediated gene transfer *in vitro* to and strong transgene expression of *LacZ* was detected in both Sertoli and Leydig cells except the germ cells following administration of Ad vectors into the adult rat testes by intratesticular injection or testis-mediated gene transfer (TMGT) [30]. Other *in vivo* and *in vitro* attempts have not yet provided evidence of infection of spermatozoa or spermatogenic cells when the Ad vectors were directly injected into mice testes [31, 32]. To date, only one study by Farre *et al.* reports that the exogenous genes can be delivered into pig spermatozoa and to a limited extent into offspring after artificial insemination (AI) or *in vitro* fertilisation if spermatozoa were incubated with Ad vectors *in vitro* [33]. In 2007, Takehashi *et al.* successfully infected spermatogonial stem cells from ROSA26 Cre reporter mice with a *Cre*-expressing Ad vector *in vitro*, and the infected spermatogonial stem cells were found to reinitiate spermatogenesis after transplantation into seminiferous tubules of infertile recipient testis [16]. However, they did not obtain transgenic offspring. A more recent study by intra-testicular injection of adenovirus expressing a green fluorescent protein (GFP) transgene also results in expression of GFP only in Sertoli cell [34]. This result is in agreement with the reports in most previous studies, and demonstrates that adenoviral infection can cause alteration in Sertoli cell function resulting in disturbances in seminiferous tubule structure and germ cell survival [34]. Gordon suggested that germline integration is likely to be less frequent with Ad vectors than with other methods, although it is possible to deliver exogenous genes into the germline with Ad vectors [35].

At this moment, data from the most studies suggest that Ad vectors can hardly promote uptake of exogenous DNA or RNA by spermatozoa, much less deliver them to the next generation. Thus Ad vectors are not likely usable for SMGT approaches involving direct incubation of sperm cells with vector.

Retroviral Vectors

Retroviruses are the one of mainstays of current gene therapy approaches, and the biology of retroviruses has been relatively well understood. They contain a reverse transcriptase which allows integration into the host genome. Most retroviral vectors are derived from the γ–retroviral genus, and called oncorectroviral vectors or simple retroviral vectors. In those vectors, there is only a simple *gag-pol-env* genome structure. Although retroviral vectors require mitotic cell division for transduction, they have a number of advantages for transgenesis and gene therapy: high efficiency of integration and gene transfer and easy production of vectors [36].

In 1985, two independent studies showed that exogenous genes could be delivered into germline cells by retroviral vectors [37, 38]. Later, Nagano *et al.* demonstrated that retroviral vectors could transfer exogenous genes into spermatogonial stem cells. The authors transduced spermatogonial stem cells with retroviral vectors either *in vitro* or *in vivo* during the transplantation process, and subsequently reestablished spermatogenesis that was maintained for six months in recipient seminiferous tubules [39]. Furthermore, their subsequent *in vitro* studies demonstrated that retroviral vectors could deliver *lacZ* gene into spermatogonial stem cells of both adult and immature mice resulted in stable integration and expression of *lacZ* gene in 2%-20% of stem cells [40]. A similar experiment was also successfully conducted in rats [41]. After carefully investigating the factors influencing retroviral transduction of spermatogonia, including the proliferative status of the infected cell, the type of viral envelope, the type of retroviral long terminal repeat, and the method of viral delivery, De Miguel and Donovan demonstrated (a) that many of the widely used retroviral vectors can be used to successfully transduce spermatogonia at high efficiency, and (b) the usefulness of the retroviral vectors of targeting substances of interest to the testis [15]. Kanatsu-Shinohara *et al.* transduced mouse spermatogonial stem cells *in situ* by microinjection of retrovirus into immature seminiferous tubules, and obtained transgenic offspring with an average efficiency of 2.8% after mating with normal females [42]. The authors improved the technique by microinjection of retroviral vectors into testis directly overcoming the drawback of *in vitro* transduction approach [42]. An improved retroviral vector, PLNCX2, is derived from the Moloney murine leukemia virus (MoMuLV) and has been designed for gene delivery and expression [43]. Furthermore, the more recent study in yaks has shown that PLNCX2 vector is useful for generating transgenic animal by SMGT [44].

Lentiviral Vectors

Lentiviruses are a special type of retroviruses which can infect both dividing and non-dividing cells [45], and have the intrinsic ability to integrate into host genomic DNA resulting in long-term expression of the transgenes *in vivo;* thus, lentiviral vectors hold great promise for various gene transfer applications. Because lentiviruses can cause a number of diseases in a range of species including human, they have to be disabled to be used as vehicles for gene transfer. During the past decade, the best characterised lentiviruses have been developed based on human immunodeficiency virus type 1 (HIV-1). The typical lentiviral vector system is based on separate vector and packaging structures. In the vector, the key genes involving in packaging and replication of the virus have been deleted from the viral genome to minimise the potential formation of replication-competent viruses. Therefore, the replication-defective vector can only go through the first phase of the infectious cycle once, and not produce infectious virus. A packaging/helper cell line, such as 293T, in *trans* with the missing proteins (*gag* and *pol*), is essential for production of the whole vector particles (Fig. **2**) [46].

Application of lentiviral vectors in SMGT was firstly reported by Nagano *et al.* in 2002 [47]. The authors transduced the male mouse germline stem cells by lentiviral vectors *in vitro*, and detected complete spermatogenesis when transplanted them into infertile host testes. In the same year, Hamra *et al.* reported the successful production of LacZ/EGFP transgenic rats by transferring germ stem cells transduced by a lentiviral vector into the testis of receipt males [48]. Using this procedure, approximately 50% of the pups in the F_2 generation inherited a lentiviral transgene. Another independent study demonstrated restoration of spermatogenesis by lentiviral gene transfer of Sertoli cells [49]. Recently, two studies reported successful production of transgenic rats by lentiviral transduction and transplantation of spermatogonial stem cells *in vitro* [50, 51]. To date, only two studies report introduction of lentiviral vectors into the testis directly by injection, but Sertoli cells (alone) are transduced and express the transgenes [49, 52]. Although there have been no reports confirming that lentiviral vectors can be used to deliver exogenous DNA into male germ cells *in vivo*, alternative vector systems may offer potential *in vivo* gene transfer in the future.

A third generation of HIV1-based vectors has recently become available, and the alternative non-primate-derived immunodeficiency viral vector system from other species and lentiviruses has been developed [36, 53, 54]. These current improvements, as well as likely future technique development, are likely to improve the utility of lentiviral vectors for animal transgenesis by SMGT. However, it remains to be seen whether lentiviral vectors can be developed such as to permit direct delivery of transgene sequences *via* incubation with sperm cells *in vitro*.

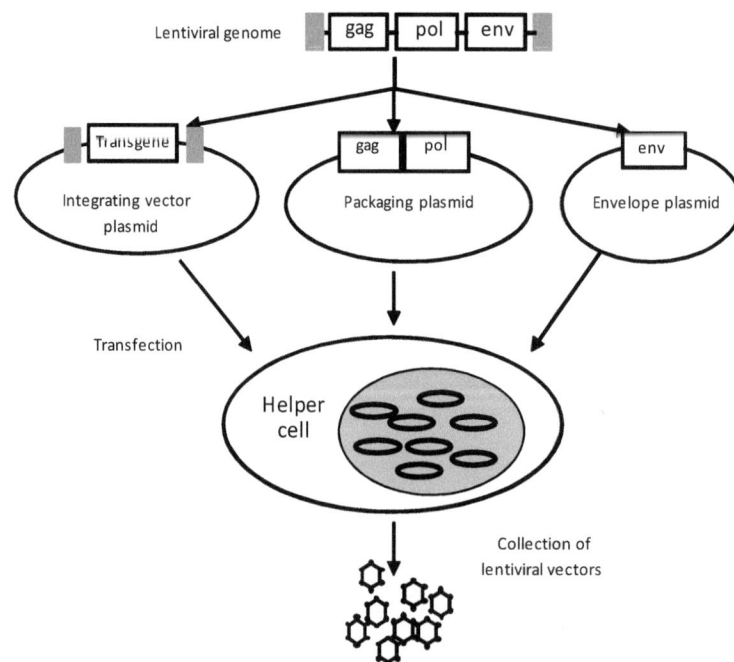

Figure 2: Production of Typical Lentiviral Vectors for Gene Transfer.

The wild-type lentiviral genome includes three main genes (*gag*, *pol*, and *env*), and long-terminal repeats (gray box). The viral genome is divided into vector (vector plasmid) and packaging parts (packaging and envelope plasmids). The vector plasmid contains the LTR (gray boxes), necessary *cis*-acting sequences and the transgene. The infectious particles are expressed and structured by transfection of vector, packaging and envelop plasmids together to the helper cells.

Other Viral Vectors

A more recent study reported that adeno-associated virus (AAV) is an efficient vector for male germ cells transduction and transgene transmission [55]. AAV is a small, nonpathogenic, dependent parvovirus with a 4.7 kb single-stranded linear genome that can integrate in a site-specific manner into human chromosome 19 in both dividing and non-dividing cells [56, 57]. Integration occurs by nonhomologous recombination at random locations if the viral *rep* is deleted. AAV is a dependent virus, so it does not carry the same biosafety restrictions as oncoretroviral or lentivral vectors, which require higher biosafety precautions for the animals. In contrast, animals transduced by AAV can be maintained under standard husbandry conditions, making this approach potentially very suitable for transgenesis in large animal species [55]. Honaramooz *et al.* exposed the male mouse germ cells collected from donor testes to AAV vectors carrying GFP transgene *in vitro*, and transplanted the modified germ cells into recipient testes. After mating of the recipient males to wild-type females, this experiment yielded 10% transgenic offspring. A similar experiment was successfully conducted in goats and 10% transgenic embryos were obtained [55]. This study demonstrates for the first time that transduction of male germ cells by AAV vectors is efficient both in mice and goats.

The Nonviral Methods

Besides viral vectors, many other techniques have been developed and improved upon during the last twenty years. In these techniques, exogenous genes are captured by spermatozoa by means of physical or chemical treatments *in vitro* or *in vivo*. This section will describe these approaches in some detail.

Autouptake

Incubation of exogenous DNA with spermatozoa, known as autouptake or simply DNA incubation, is the most direct method, reported firstly in 1971 [1]. This straightforward approach suspends the seminal plasma-free sperm cells in an appropriate medium containing exogenous DNA molecules. Then the resultant sperm cells carrying exogenous DNA are used to fertilise ova by *in vitro* fertilisation or artificial insemination (AI).

A variety of molecules or factors have been revealed to be involved during the interaction of spermatozoa and exogenous DNA, so it is logical to adjust such factors by modifying sperm cells or media in order to increase the efficiency of DNA uptake by spermatozoa. Castro *et al.* indicated that the ability of sperm cells to associate DNA molecules was affected by sperm motility and the DNA/sperm ratio [58]. Horan *et al.* further revealed that motile sperms were more efficient at capturing DNA molecules than non-motile sperms [59]. Studies have demonstrated that longer incubation times result in a higher binding rate of exogenous DNA and sperm cells followed by the internalisation of the nucleus [58, 60], but also that longer incubation time could affect sperm viability resulting in lower fertilisation rates and consequently fewer transgenic embryos [61]. Bacci *et al.* demonstrated that SMGT-treated spermatozoa retained good quality and fertilisation potential for at least 24 h, independently of the DNA dose [62].

The overall efficiency of SMGT was reported to be higher when ejaculated spermatozoa were used, compared to epididymal spermatozoa, while development of transgenic embryos was a DNA dose-dependent effect in which high DNA dose arrested embryonic development [63]. Seminal plasma contains many factors that maintain sperm motility. However, the washing step to remove the seminal plasma is desirable for capturing exogenous DNA, because there are many inhibitors affecting DNA uptake in seminal plasma [6]. Lavitrano *et al.* reportedly produced hDAF transgenic piglets at a ratio of 57% by incubation of plasmids and sperm cells after removing the seminal fluid, and transgene was expressed in 80% of transgenic piglets [64]. Kuznetsov and Kuznetsova treated sperm cells firstly using DMSO in conjunction with heat shock and successfully raised the efficiency of exogenous DNA incorporation to about 62% [13]. A high transgenic ratio (of up to 56.3%) was also obtained in subsequent studies using exogenous DNA/DMSO complex to transfect spermatozoa of mice and rabbits [65, 66].

Liposome-Mediated SMGT

Liposomes are small artificial vesicles of spherical shape that have been used for a long time as drug carriers. Liposomes can be loaded with a great variety of molecules, such as small drug molecules, proteins, nucleotides and even plasmids into living cells [67]. The liposomes carrying nucleic acid molecules are made up of cationic lipids, which can interact with the negatively charged nucleic acid molecules and form complexes coating the nucleic acid inside [11]. Such liposome-mediated gene transfer is now an established means of genetically altering cells *in vitro*. Bachiller *et al.* demonstrated that DNA transfer into sperm by liposomes was very efficient although no transgenic mice were produced [11]. However, several studies have shown that exogenous DNAs can be taken up efficiently by sperm cells and transmitted to next generation if they are coated by liposomes in advance [19, 68-71].

Yonezawa *et al.* compared eight commercially available liposomes in associating exogenous DNA with rat sperm cells, and found that only two liposomes, DMRIE-C and SuperFect™, led to detection of exogenous DNA in spermatozoa. By means of TMGT using either of the two liposomes, more than 80% of morula-stage embryos expressed EGFP transgene. Although the success rate of transgenic offspring was still limited, the authors suggested that exogenous DNA could be integrated into the genome of the progeny under specific conditions, such as selection of the liposomes [69]. A subsequent experiment incubated rabbit spermatozoa with liposome/DNA mixture before *in vitro* fertilisation, and 66% spermatozoa were detected carrying transgene. Transgene expression was found in different stages of embryos and in tissues of young rabbits [70]. Chang *et al.* demonstrated that foreign DNA injected into the testis as liposome complexes can be transferred into eggs *via* sperms and expressed in the postpartum progeny [68]. High transgenic rates were reported more recently in mouse F_1 (41%) and F_2 (37%) offspring *via* TMGT using liposome- (DOTAP) treated plasmid DNA [19]. Additional, a recent study showed that liposomes made of lipids isolated from sperm membrane, so-called "spermatosomes", could undergo strong membrane-membrane fusion, resulting in an effective transfer of foreign molecules to the cytosol of target cells [72]. This fusogenic potential of sperm membrane lipids provides a new way to make liposomes which mediate gene transfer *via* SMGT. Another alternative of using liposome in SMGT is the utilisation of liposome-peptide-DNA (LPD) complex with the aid of protamine. LPD is considered to be helpful for stabilizing the liposome-DNA complex during fertilisation, and exogenous DNA can be detected in the offspring using LPD *via* SMGT [73]. Therefore, LPD complex may be useful in transgenesis by SMGT.

Linker-Based SMGT

Linker or receptor-mediated gene transfer (linker-based SMGT) was first reported in 1987 using polycation-conjugated asialoglycoprotein [74]. DNA molecules can bind to polycations in a strong but noncovalent manner forming soluble complexes. DNA coupled with antibodies or antibody-fragments offers the ability to internalise the complexes *via* receptor-mediated endocytosis [75]. Qian *et al.* were the first to report that monoclonal antibody (mAbC) could be used as linker for SMGT to generate transgenic mice and chickens [76, 77]. Subsequently, Chang *et al.* successfully used a monoclonal antibody (mAbC) as linker to bind exogenous DNAs to sperm cells, then obtained viable pig and mouse offspring with integration of exogenous DNA into the genomes *via* artificial fertilisation [78]. The authors demonstrated that foreign DNA could be specifically bound to the sperm cell surface *via* the linker protein (mAbC) through ionic interaction. The antibody cross-linker may provide protection from DNase activity designed to prevent exogenous DNA molecules from entering the eggs. Current data indicate that linker-based SMGT can be used to generate transgenic animals efficiently in many different species, especially in the farm livestock.

Electroporation

Electroporation (EP) is valid for obtaining stable transformants in eukaryote cell lines and introducing plasmids into bacteria. EP has been used to improve uptake of exogenous DNA by sperm cells since 1991, when Gagné *et al.* treated bovine spermatozoa *in vitro* by electrical field initially with several combinations of capacitance and voltage [79]. They observed that electroporated sperm cells retained DNA molecules more efficiently than spermatozoa not submitted to the electrical field. Their study demonstrated that foreign DNA could be stably captured by spermatozoa following EP treatment, to be carried into oocytes during fertilisation. However, it is still requisite to transplant embryos so-obtained.

Interaction of Sperm Cells and Exogenous DNA In Vivo

Capture of exogenous DNA by sperm cells *in vivo* has been reported in transgenesis since 1994, when Sato *et al.* first attempted to transfect mice testicular spermatozoa and spermatogonia by direct injection of naked plasmid DNA [18]. Although they did not obtain transgenic animals, their pioneering work opens a new way for animal transgenesis. This *in vivo* technology introduces exogenous DNA directly into testis by injection, as a form of nonviral testis-mediated gene transfer (TMGT).

In TMGT, interaction of sperm cells and exogenous DNA occurs *in vivo*, and the methods above mentioned to improve capture interactions are also practical. Choice of reagents used for gene transfer is important for improvement of TMGT. For example, use of DMRIE-C, TransFect™, and DOTAP appears to be promising [19, 69]. Ogawa *et al.* demonstrated that repeated injection of linearised plasmid DNA encapsulated with cationic liposome into testis resulted in transmission of exogenous DNA to F_0 progeny (blastocysts) by fertilisation [80]. Sato *et al.* showed that a single injection of circular plasmid DNA encapsulated with liposome into mature mouse testis was sufficient for transfection of sperm cells, and a high ratio of gene delivery to F_0 generation (mid-gestational fetuses) was obtained by natural mating of injected males with normal females [81]. A further study demonstrated that testicular spermatozoa could be transfected by linear plasmid DNA, and the exogenous DNA could be transmitted from F_0 to F_2 transgenic mice by natural mating, if the plasmid DNA was initially mixed with nonliposomal lipid FuGENE™ 6 Transfection Reagent [82]. Repeated injections were not critical to introduction of high numbers of copies of DNA into eggs [83]. Sato *et al.* suggested that the pattern of transmission of exogenous DNA to the next generation may not be Mendelian [82]. Furthermore, current studies suggest that exogenous DNA is extensively lost during transition from the preimplantation to postimplantation stages [69, 82, 83].

Many studies showed that *in vivo* EP was an efficient method [84, 85]. Efficient and convenient expression of transgenes during spermatogenesis was obtained by injection of exogenous DNA into seminiferous tubules with subsequent *in vivo* electroporation using an electroporator [86]. Hibbitt *et al.* injected an expression vector into the rete testis of hamster, then an electrical current was applied to the testis using a voltage of 50 V and pulse length of 50 ms. The authors provided the first systematic demonstration that EP did not lead to any significant long-term adverse effects on testicular integrity and sperm quality [87]. Interestingly, the rete testis was indicated as the suitable site for injection for *in vivo* gene transfer by EP [88].

Introduction of exogenous DNA into the testis of a very young male may be a convenient option. Because young testes have actively proliferating spermatogonial cells, the sperm cells may be easily accessible for exogenous DNA contact [89].

Magnetofection

Magnetofection technique has been studied extensively for the purpose of delivering transgenes into target cells or tissues [90, 91]. The basic principle of magnetofection is to associate magnetic nanoparticles (MNPs) with transgene molecules, so that the complexes can be delivered into intracellular spaces using a magnetic field (Fig. **3**). Recently, Kim *et al.* reported the successful use of MNPs to efficiently introduce a transgene, pCX-EGFP/Neo, into the embryo *via* spermatozoa [17].

DNA incubation is the straightforward method used in SMGT. If the magnetofection technique is employed, the dishes will be placed under a magnetic field (A). Alternative means of the direct method include combination of sperm/DNA interaction with helpers, such as liposome, linker, and acceptor molecules (B), and utilisation of virus vectors carrying the foreign DNA (C). Oocytes are subsequently fertilised by either IVF or ICSI. The pseudopregnant female can also be fertilised by AI directly. The viable embryos are selected and implanted into the pseudopregnant females. TMGT can be performed by direct injection within testis with spermatogonial stem cells carrying foreign DNA molecules (D). DNA, DNA-liposome or virus vector can also be introduced into sperm cells by direct injection within the seminiferous tubules or within the rete testis *in vivo*, and electroporation can be performed to promote entry of the foreign DNA into sperm cells *in situ* or *in vivo* (E). (See detail in Ref [92])

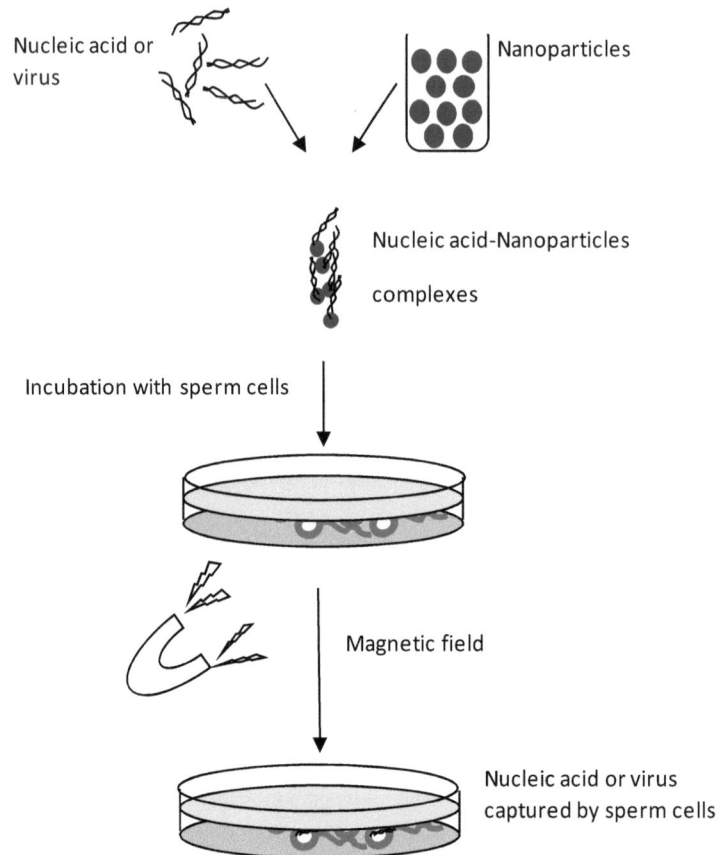

Figure 3: Schematic Procedure of Magnetofection Technique used for Producing Transgenesis by SMGT. Magnetic nanoparticles coated with cationic molecules are used to associate with DNA or virus. The DNA-nanoparticles and the sperm cells are incubated under a magnetic field, and the DNA or virus can be captured by sperm cells.

METHODS FOR PRODUCING TRANSGENIC ANIMALS AFTER CAPTURE OF EXOGENOUS DNA BY SPERM CELLS

Besides the important procedures employed to facilitate the capture exogenous DNA by sperm cells (as described above), the key step required to produce transgenic animals is to enable the modified sperms to fertilise eggs. During the past two decades, several of the well-developed fertilisation techniques employed to treat infertility or for biotechnological use, such as *in vitro* fertilisation (IVF), intracytoplasmic sperm injection (ICSI), artificial insemination (AI), have been successfully used with SMGT. Gamete intrafallopian transfer (GIFT) is also of potential use in SMGT. Finally, natural mating, following TMGT, is the most simple and reliable method to produce transgenic offspring.

In Vitro Fertilisation (IVF)

In vitro fertilisation is commonly referred to as IVF. IVF in SMGT is the process of fertilisation by manually combining an egg and modified sperms carrying transgenes in a laboratory dish. When the IVF procedure is successful, the process is combined with a procedure known as embryo transfer, which is used to physically place the embryo in the uterus of pseudo-pregnant mothers (Fig. **4**).

Intracytoplasmic Sperm Injection (ICSI)

Intracytoplasmic sperm injection (ICSI) is an *in vitro* fertilisation procedure in which a single sperm is injected directly into an egg (Fig. **4**). The procedure is done under a microscope using multiple micromanipulation devices (micromanipulator, microinjectors and micropipettes). A holding pipette stabilises the mature oocyte with gentle suction applied by a microinjector. From the opposite side a thin, hollow glass micropipette is used to collect a single sperm carrying exogenous DNA, having immobilised it by cutting its tail with the point of the micropipette. The micropipette is pierced through the oolemma and into the inner part of the oocyte (cytoplasm). The sperm is then released into the oocyte. After the procedure, the oocyte will be placed into cell culture and checked on the following day for signs of fertilisation. The embryos will be cultured in mediums or transplanted into the uterus of pseudo-pregnant mothers.

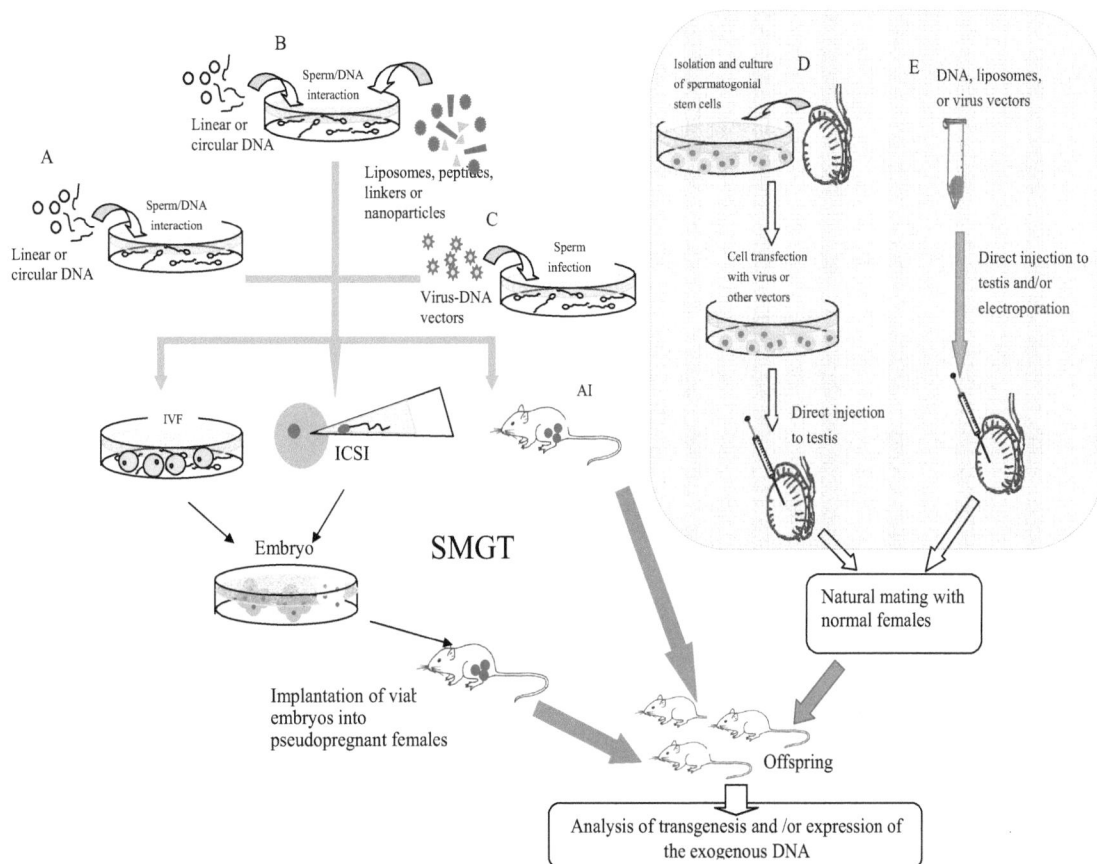

Figure 4: Schematic Representation of a Variety of Methods used for Producing Transgenesis by SMGT and TMGT.

Artificial Insemination (AI) and Gamete Intrafallopin Transfer (GIFT)

The procedure of artificial insemination (AI) is to place sperm (or semen) treated with transgenes in the reproductive tract of a female with the intention of impregnating the female (Fig. **4**). The main techniques used in AI include intracervical insemination (ICI) and intrauterine insemination (IUI). ICI is the easiest way to inseminate, where modified sperm and medium are injected high into the cervix with a needle-less syringe. By contrast, IUI is suitable for washed sperms removed from most other components of the seminal fluid. In IUI, the modified sperms can be injected directly into the female uterus.

The method of gamete intrafallopian transfer (GIFT) is a tool of assisted reproductive technology against infertility. It is a variation of IVF. Oocytes are removed from female's ovaries, and placed in one of the fallopian tubes, along with the sperms carrying exogenous DNAs. The technique allows fertilisation to take place inside the female's body.

Natural Mating

As discussed above, TMGT is an alternative means of SMGT. This *in vivo* technology introduces foreign DNA directly into testis by injection and promises mixture of exogenous DNA and sperm cells in testis. The treated males will deliver the transgenes to next generations by natural mating with normal (untreated) females (Fig. **4**).

REFERENCES

[1]   Brackett BG, Baranska W, Sawicki W, *et al.* Uptake of heterologous genome by mammalian spermatozoa and its transfer to ova through fertilisation. Proc Natl Acad Sci USA 1971; 68: 353-57.

[2]   Arezzo F. Sea urchin sperm as a vector of foreign genetic information. Cell Biol Int Rep 1989; 13: 391-04.

[3]   Lavitrano M, Camaioni A, Fazio VM, *et al.* Sperm cells as vectors for introducing foreign DNA into eggs: genetic transformation of mice. Cell 1989; 57: 717-23.

[4]   Brinster RL, Sandgren EP, Behringer RR, *et al.* No simple solution for making transgenic mice. Cell 1989; 59: 239-41.

[5]   Birnstiel ML, Busslinger M. Dangerous liaisons: spermatozoa as natural vectors for foreign DNA? Cell 1989; 57: 701-02.

[6]   Lavitrano M, French D, Zani M, *et al.* The interaction between exogenous DNA and sperm cells. Mol Reprod Dev 1992; 31: 161-69.

[7]   Giordano R, Magnano AR, Zaccagnini G, *et al.* Reverse transcriptase activity in mature spermatozoa of mouse. J Cell Biol 2000; 148: 1107-13.

[8]   Zoraqi G, Spadafora C. Integration of foreign DNA sequences into mouse sperm genome. DNA Cell Biol 1997; 16: 291-00.

[9]   Webster N, Forni M, Bacci ML, *et al.* Multi-transgenic pigs expressing three fluorescent proteins produced with high efficiency by sperm mediated gene transfer. Mol Reprod Dev 2005; 72: 68-76.

[10]   Zani M, Lavitrano M, French D, *et al.* The mechanism of binding of exogenous DNA to sperm cells: factors controlling the DNA uptake. Exp Cell Res 1995; 217: 57-64.

[11]   Bachiller D, Schellander K, Peli J, *et al.* Liposome mediated DNA uptake by sperm cells. Mol Reprod Dev 1991; 30: 194-00.

[12]   Gagne MB, Pothier F, Sirard MA. Electroporation of bovine spermatozoa to carry foreign DNA in oocytes. Mol Reprod Dev 1991; 29: 6-15.

[13]   Kuznetsov AV, Kuznetsova IV. The binding of exogenous DNA pRK31acZ by rabbit spermatozoa, its transfer to oocytes and expression in preimplantation embryos. Ontogenez 1995; 26: 300-09.

[14]   Perry ACF, Wakayama T, Kishikawa H, *et al.* Mammalian transgenesis by intracytoplasmic sperm injection. Science 1999; 284: 1180-83.

[15]   De Miguel MP, Donovan PJ. Determinants of retroviral-mediated gene delivery to mouse spermatogonia. Biol Reprod 2003; 68: 860-66.

[16]   Takehashi M, Kanatsu-Shinohara M, Inoue K, *et al.* Adenovirus-mediated gene delivery into mouse spermatogonial stem cells. Proc Natl Acad Sci USA 2007; 104: 2596-01.

[17]   Kim TS, Lee SH, Gang GT, *et al.* Exogenous DNA uptake of boar spermatozoa by a magnetic nanoparticle vector system. Reprod Dom Anim 2009; Online 10.1111/j.1439-0531.2009.01516.x

[18]   Sato M, Iwase R, Kasai K, *et al.* Direct injection of foreign DNA into mouse testis as a possible alternative of sperm-mediated gene transfer. Anim Biotechnol 1994; 5: 19-31.

[19]   He X, Qi B, Liu G, *et al.* A novel method to transfer gene *in vivo* system. Prog Biochem Biophys 2006; 33: 685-90.

[20]   Kurita K, Burgess SM, Sakai N. Transgenic zebrafish produced by retroviral infection of *in vitro*-cultured sperm. Proc Natl Acad Sci USA 2004; 101: 1263-67.

[21]   Jonak J. Sperm-mediated preparation of transgenic *Xenopus laevis* and transgenic DNA to the next generation. Mol Reprod Dev 2000; 56: 298-00.

[22]   Nakanishi A, Iritani A. Gene transfer in the chicken by sperm-mediated methods. Mol Reprod Dev 1993; 36: 258-61.

[23]   Yin J, Zhang JJ, Shi GG, *et al.* Sperm mediated human coagulation factor VIII gene transfer and expression in transgenic mice. Swiss Med Wkly 2009; 139: 364-72.

[24]   Chen HL, Yang HS, Huang R, *et al.* Transfer of a foreign gene to Japanese abalone (*Haliotis diversicolor supertexta*) by direct testis-injection. Aquaculture 2006; 253: 249-58.

[25]   Gao DS, Sun LX, Zeng ZJ, *et al.* Sperm-Mediated Gene Transfer in the Chinese Honeybee, *Apis cerana cerana* (Hymenoptera: Apidae). J Asia-Pacific Entomol 2007; 10: 339-44.

[26]   Danthinne X, Imperiale MJ. Production of first generation adenovirus vectors: a review. Gene Ther 2000; 7: 1707-14.

[27]   Fallaux FJ, Kranenburg O, Cramer SJ, *et al.* Characterisation of 911: a new helper cell line for the titration and propagation of early region 1-deleted adenoviral vectors. Hum Gene Ther 1996; 7: 215-22.

[28]   Murakami P, Pungor E, Files J, *et al.* A single short stretch of homology between adenoviral vector and packaging cell line can give rise to cytopathic effect-inducing, helper-dependent E1-positive particles. Hum Gene Ther 2002; 13: 909-20.

[29]   Hitt M, Bett A, Addison C, *et al.* Techniques for human adenovirus vector construction and characterisation. Methods Mol Genet 1995; 7: 13-30.

[30]   Blanchard KT, Boekelheide K. Adenovirus-mediated gene transfer to rat testis *in vivo.* Biol Reprod 1997; 56: 495-00.

[31]   Kanatsu-Shinohara M, Ogura A, Ikegawa M, *et al.* Adenovirus-mediated gene delivery and *in vitro* microinsemination produce offspring from infertile male mice. Proc Natl Acad Sci USA 2002; 99: 1383-88.

[32]   Kojima Y, Sasaki S, Umemoto Y, *et al.* Effects of adenovirus mediated gene transfer to mouse testis *in vivo* on spermatogenesis and next generation. J Urol 2003; 170: 2109-14.

[33]   Farre L, Rigau T, Mogas T, *et al.* Adenovirus-mediated introduction of DNA into pig sperm and offspring. Mol Reprod Dev 1999; 53: 149-58.

[34]   Hooley RP, Paterson M, Brown P, *et al.* Intra-testicular injection of adenoviral constructs results in Sertoli cell-specific gene expression and disruption of the seminiferous epithelium. Reproduction 2009; 137: 361-70.

[35]   Gordon JW. Direct exposure of mouse ovaries and oocytes to high doses of an adenovirus gene therapy vector fails to lead to germ cell transduction. Mol Ther 2001; 3: 557-64.

[36]   Friedmann T, Rossi J. Gene transfer: delivery and expression of DNA and RNA. New York: Cold Spring Harbor; 2007.

[37]   van der Putten H, Botteri FM, Miller AD, *et al.* Efficient insertion of genes into the mouse germline *via* retroviral vectors. Proc Natl Acad Sci USA 1985; 82: 6148-52.

[38]   Jahner D, Haase K, Mulligan R, *et al.* Insertion of the bacterial gpt gene into the germline of mice by retroviral infection. Proc Natl Acad Sci USA 1985; 82: 6927-31.

[39]   Nagano M, Shinohara T, Avarbock MA, *et al.* Retrovirus-mediated gene delivery into male germline stem cells. FEBS Lett 2000; 475: 7-10.

[40]   Nagano M, Brinster CJ, Orwig KE, *et al.* Transgenic mice produced by retroviral transduction of male germ-line stem cells. Proc Natl Acad Sci USA 2001; 98: 13090-95.

[41]   Orwig KE, Avarbock MR, Brinster RL. Retrovirus-mediated modification of male germline stem cells in rats. Biol Reprod 2002; 67: 874-79.

[42]   Kanatsu-Shinohara M, Toyokuni S, Shinohara T. Transgenic mice produced by retroviral transduction of male germline stem cells *in vivo.* Biol Reprod 2004; 71: 1202-07.

[43]   Miller AD, Rosman GJ. Improved retroviral vectors for gene transfer and expression. BioTechniques 1989; 7: 980-90.

[44]   Zi XD, Chen SW, Liang GN, *et al.* The effect of retroviral vector on uptake of human lactoferrin DNA by yak (*Bos grunniens*) spermatozoa and their fertilizability *in vitro.* Anim Biotech 2009; 20: 247-51.

[45]   Naldini L, Blomer U, Gallay P, *et al. In vivo* gene delivery and stable transduction of nondividing cells by a lentiviral vector. Science 1996; 272: 263-67.

[46]   Pfeifer A. Lentiviral transgenesis. Transgenic Res 2004; 13: 513-22.

[47]   Nagano M, Watson DJ, Ryu BY, *et al.* Lentiviral vector transduction of male germline stem cells in mice. FEBS Lett 2002; 524: 111-15.

[48]   Hamra FK, Gatlin J, Chapman KM, *et al.* Production of transgenic rats by lentiviral transduction of male germ-line sten cells. Proc Natl Acad Sci USA 2002; 99: 14931-36.

[49]   Ikawa M, Tergaonkar V, Ogura A, *et al.* Restoration of spermatogenesis by lentiviral gene transfer: offspring from infertile mice. Proc Natl Acad Sci USA 2002; 99: 7524-29.

[50]   Ryu BY, Orwig K, Oatley JM, *et al.* Efficient generation of transgenic rats through the male germline using lentiviral tansduction and tansplantation of spermatogonial stem cells. J Androl 2007; 28: 353-60.

[51]   Kanatsu-Shinohara M, Kato M, Takehashi M, *et al.* Production of transgenic rats *via* lentiviral transduction and xenogeneic transplantation of spermatogonial stem cells. Biol Reprod 2008; 79: 1121-28.

[52]   Wang F, Zhang Q, Cao J, *et al.* The microtubule plus endbinding protein EB1 is involved in Sertoli cell plasticity in testicular seminiferous tubules. Exp Cell Res 2008; 314: 213-26.

[53]   McGrew MJ, Sherman A, Ellard FM, *et al.* Efficient production of germline transgenic chickens using lentiviral vectors. EMBO Rep 2004; 5: 728-33.

[54]   Poeschla EM. Non-primate lentiviral vectors. Curr Opin Mol Ther 2003; 5: 529-40.

[55]   Honaramooz A, Megee S, Zeng W, *et al.* Adeno-associated virus (AAV)-mediated transduction of male germline stem cells results in transgene transmission after germ cell transplantation. FASEB J 2008; 22: 374-82.

[56]   Russell DW, Kay MA. Adeno-assocoated virus vectors and hematology. Blood 1999; 94: 864-74.

[57]   Srivastava A. Obstacles to human hematopoietic stem cell transduction by recombinant adeno-associated virus 2 vectors. J Cell Biochem 2002; 85: 39-45.

[58]   Castro FO, Hernández O, Uliver C, *et al.* Introduction of foreign into the spermatozoa of farm animals. Theriogenology 1990; 34: 1099-10.

[59]   Horan R, Powell R, Mcquaid S, *et al.* Association of foreign DNA with porcine spermatozoa. Arch Androl 1991; 26: 83-92.

[60]   Francolini M, Lavitrano M, Lamia CL, *et al.* Evidence for nuclear internalisation of exogenous DNA into mammalian sperm cells. Mol Reprod Dev 1993; 34: 133-39.

[61]   Feitosa WB, Milazzotto MP, Simões R, *et al.* Bovine sperm cells viability during incubation with or without exogenous DNA. Zygote 2009; 17: 315-20.

[62]   Bacci ML, Zannoni A, De Cecco M, *et al.* Sperm-mediated gene transfer–treated spermatozoa maintain good quality parameters and *in vitro* fertilisation ability in swine. Theriogenology 2009; 72: 1163-70.

[63]   Sciamanna I, Piccoli S, Barberi L, *et al.* DNA dose and sequence dependence in sperm-mediated gene transfer. Mol Reprod Dev 2000; 56: 301-305.

[64]   Lavitrano M, Forni M, Bacci ML, *et al.* Sperm mediated gene transfer in pigs: selection of donor boars and optimisation of DNA uptake. Mol Reprod Dev 2003; 64: 284-91.

[65]   Li L, Shen W, Min L, *et al.* Human *lactoferrin* transgenic rabbits produced efficiently using dimethylsulfoxide–sperm-mediated gene transfer. Reprod Fertil Dev 2006; 18: 689-95.

[66]   Shen W, Li L, Pan Q, *et al.* Efficient and simple production of transgenic mice and rabbits using the new DMSO-sperm mediated exogenous DNA transfer method. Mol Reprod Dev 2006; 73: 589-94.

[67]   Gregoriadis G, Allison AC. Liposomes in Biological Systems. New York: John Wiley and Sons; 1980.

[68]   Chang KT, Ikeda A, Hayashi K, *et al.* Production of transgenic rats and mice by the testis-mediated gene transfer. J Reprod Dev 1999; 45: 29-36.

[69]   Yonezawa T, Furuhata Y, Hirabayashi K, *et al.* Detection of transgene in progeny at different developmental stages following testis-mediated gene transfer. Mol Reprod Dev 2001; 60: 196-01.

[70]   Wang HJ, Lin AX, Chen YF. Association of rabbit sperm cells with exogenous DNA. Anim Biotechnol 2003; 14: 155-65.

[71]   Ball BA, Sabeur K, Allen WR. Liposome-mediated uptake of exogenous DNA by equine spermatozoa and applications in sperm-mediated gene transfer. Equine Vet J 2008; 40: 76-82.

[72]   Atif SM, Hasan I, Ahmad N, *et al.* Fusogenic potential of sperm membrane lipids: Nature's wisdom to accomplish targeted gene delivery. FEBS Lett 2006; 580: 2183-90.

[73]   Yonezawa T, Furuhata Y, Hirabayashi K, *et al.* Protamine-derived synthetic peptide enhances the efficiency of sperm-mediated gene transfer using liposome-peptide-DNA complex. J Reprod Dev 2002; 48: 281-86.

[74]   Wu GY, Wu CH. Receptor-mediated *in vitro* gene transformation by a soluble DNA carrier system. J Biol Chem 1987; 262: 4429-32.

[75]   Varga CM, Wickham TJ, Lauffenburger DA. Receptor-mediated targeting of gene delivery vectors: Insights from molecular mechanisms for improved vehicle design. Biotechnol Bioeng 2000; 70: 593-05.

[76]   Qian J, Chang K, Jiang M, *et al.* Generation of transgenic pigs by sperm-mediated gene transfer using a linker protein (mAbc). Abstracts of the 3rd UC Davis Transgenic Animal Research Conference, Tahoe City, California. The Netherlands: Kluwer Academic Publishers; 2001.

[77]   Qian J, Liu Y, Jiang M, *et al.* A novel and highly effective method to generate transgenic mice and chickens: linker based sperm-mediated gene transfer. Abstracts of the 3rd UC Davis Transgenic Animal Research Conference, Tahoe City, California. The Netherlands: Kluwer Academic Publishers; 2001.

[78]   Chang K, Qian J, Jiang MS, *et al.* Effective generation of transgenic pigs and mice by linker based sperm-mediated gene transfer. BMC Biotechnol 2002; 2: 5-17.

[79]   Gagné MB, Pothier F, Sirard MA. Electroporation of bovine spermatozoa to carry foreign DNA in oocytes. Mol Reprod Dev 1991; 29: 6-15.

[80]   Ogawa S, Hayashi K, Tada N, *et al.* Gene expression in blastocysts following direct injection of DNA into testis. J Reprod Dev 1995; 41: 379-82.

[81] Sato M, Gotoh K, Kimura M. Sperm-mediated gene transfer by direct injection of foreign DNA into mouse testis. Transgenics 1999; 2: 357-69.

[82] Sato M, Yabuki K, Watanabe T, *et al.* Testis-mediated gene transfer (TMGT) in mice: successful transmission of introduced DNA from F0 to F2 generation. Transgenics 1999; 3: 11-22.

[83] Sato M, Nakamura S. Testis-mediated gene transfer (TMGT) in mice: effects of repeated DNA injections on the efficiency of gene delivery and expression. Transgenics 2004; 4: 101-19.

[84] Heller R, Jaroszeski M, Atkin A, *et al. In vivo* gene electroinjection and expression in rat liver. FEBS Lett 1996; 389: 225-28.

[85] Nishi T, Yoshizato K, Yamashiro S, *et al.* High-efficiency *in vivo* gene transfer using intraarterial plasmid DNA injection following *in vivo* eletroporation. Cancer Res 1996; 56: 1050-55.

[86] Yamazaki Y, Fujimoto H, Ando H, *et al. In vivo* gene transfer to mouse spermatogenic cells by deoxyribonucleic acid injection into seminiferous tubules and subsequent electroporation. Biol Reprod 1998; 59: 1439-44.

[87] Hibbitt O, Coward K, Kubota H, *et al. In vivo* gene transfer by elctroporation allows expression of a fluorescent transgene in hamster testis and epididymal sperm and has no adverse effects upon testicular integrity or sperm quality. Biol Reprod 2006; 74: 95-01.

[88] Kubota H, Hayashi Y, Kubota Y, *et al.* Comparison of two methods of *in vivo* gene transfer by electroporation. Fertil Steril 2005; 83: 1310-18.

[89] Sato M. Transgenesis *via* sperm. J Mamm Ova Res 2005; 22: 92-100.

[90] Plank C, Schillinger U, Scherer F, *et al.* The magnetofection method: using magnetic force to enhance gene delivery. Biol Chem 2003; 384: 737-47.

[91] Scherer F, Anton M, Schillinger U, *et al.* Magnetofection: enhancing and targeting gene delivery by magnetic force *in vitro* and *in vivo*. Gene Ther 2002; 9: 102-109.

[92] Niu Y, Liang S. Progress in gene transfer by germ cells in mammals. J Genet Genomics 2008; 35: 701-14.

CHAPTER 6

Murine SMGT: An Overview of Research

Xiaofeng Sun[*] and Wei Shen

Qingdao Agricultural University, China

Abstract: Sperm-mediated gene transfer (SMGT) in mice has undergone rapid developments over the past twenty years, because of the obvious advantages in operation: it is inexpensive and convenient, and requires minimal specialist training and equipment. The typical method of SMGT technology is to use the mature spermatozoa as vehicles for DNA delivery *in vitro*. However, the low efficiency associated with this approach has resulted in many negative results, with consequent controversy over its biological basis. By contrast, *in vivo* gene transfer through direct introduction of foreign DNA into the male reproductive tract is developing apace. The mouse as a major model animal has made an important contribution to SMGT progress.

Keywords: Murine SMGT, SCNT, Electroporation, DNA-binding proteins, SMGT, Transgenesis, DNA integration, Virus-SMGT, ICSI, TMGT.

INTRODUCTION

Transgenic mice provide a powerful approach for studying gene regulation *in vivo*, and have found widespread use as animal models of human disease and experimental systems for the production of genetically modified substances and tissues.

There are several methods for transgenesis, such as pronuclear microinjection (PNI) [1] and somatic cell nuclear transfer (SCNT) [2]. PNI involves the microinjection of transgenic DNA into the pronucleus of a fertilised egg and leads to the random insertion of exogenous DNA into the genome. This approach in most mammals, have been successful, with transgenic offspring obtained with high efficiency, short cycle time and unlimited size of transgenes. But several features greatly limit the application of PNI, including: high level skills required for embryo manipulation; low embryo survival rate; random integration of the transgene; copy number variability; and high cost [3, 4]. With SCNT, low efficiency is a major obstacle to widespread use of this technology [5]. Many factors such as donor cell type and donor cell stage, genetic identity, passage number and preparation methods of somatic donor can influence successful birth of live SCNT offspring.

An alternative method for producing transgenic animals, namely sperm-mediated gene transfer (SMGT), has been developed [6]. Its simplicity and minimal embryo manipulation provide advantages over PNI and SCNT in generating transgenic animals [7]. SMGT in vertebrates has undergone various developments over the last twenty years. The obvious advantages in operation of SMGT, particularly its lack of expense and high convenience, make it potentially very attractive. Thus SMGT has been examined by various scientists worldwide, and represents additional possibilities for existing transgenic technology. In 1989, Lavitrano reported the first successful transgenic mice generated by *in vitro* fertilisation of spermatozoa incubated with exogenous DNA [8]. While several efforts to replicate the procedure failed [9], others showed successful transfer of transgenic constructs into mice (and, in addition, several larger animal types). Similarly, transgenic mice were efficiently generated by various enhanced forms of SMGT, including a sperm-binding monoclonal antibody-based form of SMGT [10]. It appears that mammalian sperm cells are capable of binding and transferring exogenous DNA into oocytes at fertilisation by *in vitro* fertilisation or artificial insemination [11, 12]. Yang *et al.* have tested SMGT protocols with myostatin propeptide transgene. The transgene construct has been shown to significantly enhance the muscle growth in mice [13, 14]. To increase the efficiency of fertilisation, deep post-cervical intrauterine insemination of the sperm

*Address correspondence to Xiaofeng Sun: Laboratory of Germ Cell Biology, Qingdao Agricultural University, Qingdao 266109, China; Tel: 86-0532-88030246; E-mail: sunxf2008@hotmail.com

was employed, after *in vitro* incubations with foreign DNA. It showed that the ability of sperm uptaking DNA was highly correlated with sperm motility at the time of collection by *in vitro* incubation of radioactively labeled DNA. Transgenic animals are usually achieved by co-incubating the spermatozoa with the transgenes, by liposome transformation or by electroporation [15].

The process of SMGT typically proceeds as follows (as summarized in Fig. **1**): sperm are isolated from caudae epididymides of male animals; seminal fluid is removed by washing sperm in culture medium; sperm is incubated with, or transfected with, foreign DNA; the GM sperm is subsequently used to fertilise female animal *via* artificial insemination.

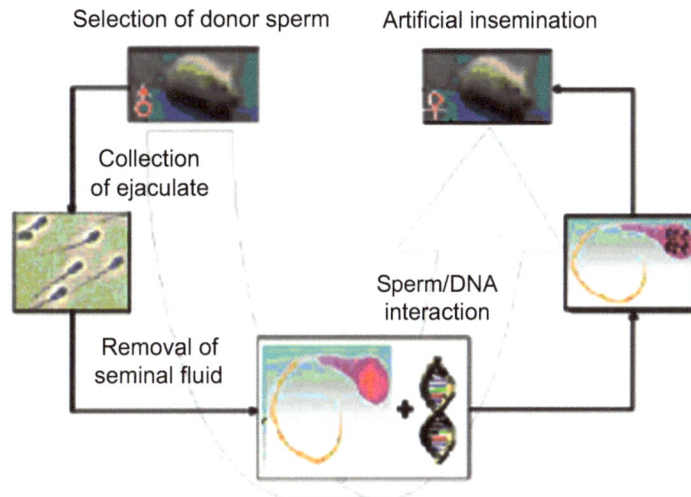

Figure 1: Murine SMGT.

MECHANISM OF SPERM MEDIATED GENE TRANSFER

Sperm cells from a variety of species share the spontaneous ability to bind and integrate foreign DNA [16]. Transgenes can indeed enter the sperm cells and can be expressed after fertilisation. However, the fields of molecular biology and *in vitro* fertilisation (IVF) were at that time still at a relatively primitive stage; accordingly, the phenomenon did not attract people's attention. In 1989, Lavitrano *et al.* published a ground-breaking SMGT paper. The Lavitrano group incubated mouse epididymal sperm and Linear Plasmid pSV2cat carrying the CTA gene, and acquired 30% transgenic offspring mice after *in vitro* fertilisation and transplantation of 2-cell embryos into the host mouse uterus. Thereafter, many successes of transgenic research by this method have been reported, with many kinds of transgenic animals offspring obtained, which has further proved the ability of sperm to bind and transfer transgenes. The sperm-DNA binding is not random; DNA usually binds to the postacrosomal region of mouse sperm head. Within this region, there is a preferred DNA binding site which can bind negatively charged macromolecules, such as DNA and some proteins (isoelectric point <7). The binding is mediated by specific 30~35KD DNA-binding proteins (DBP) which is antagonized by an inhibitory factor (IF) in the seminal fluid. So mature murine spermatozoa are naturally protected against the intrusion of foreign nucleic acid molecules, thus maintaining the genetic stability of species. However, when sperm are washed fully, or its membrane disrupted, the antagonistic function of IF is removed. A portion of sperm-bound DNA is internalized in nuclei, and the process is mediated by CD4 molecules. Sperm interaction with foreign DNA triggers endogenous nuclease (s) that cleaves both the exogenous and the genomic DNA, eventually leading to a cell death process which resembles apoptosis. Internalized foreign DNA sequences reach the nuclear matrix and may undergo recombination with chromosomal DNA. The specific 30~35KD DBP exist from lower animals to mammalians, the sequence of the amino acid is conserved among species. From these studies, a surprising network of metabolic functions is beginning to emerge in mature spermatozoa, which are normally repressed and are specifically activated upon exposure to appropriate stimuli [17]. This is the molecular model of SMGT of Spadafora and Sciamanna, as described in Chapter 12 of this book.

DIRECT METHODS OF MURINE SMGT

Directly Incubating Foreign DNA with Sperm

The simplest form of SMGT is directly incubating sperm cells with transgenes, followed by fertilisation to produce transgenic offspring. The experiments of several different laboratories have demonstrated that both circular and linearised plasmids can integrate into the genome of sperm following simple co-incubation. For example, Maione incubated mouse sperm with plasmid DNA during the capacitation period and then added the treated sperm to freshly ovulated mouse oocytes for fertilisation. Cleaved embryos were then transferred to the oviduct of pseudopregnant recipient mice for gestation. From a total of 75 experiments, 13 produced 130 transgenic offspring, amounting to 7.4% of total fetuses. In five experiments, more than 85% of offspring were transgenic [18]. The results from several groups in general show that mammalian sperm cells are capable of binding and transferring exogenous DNA into oocytes at fertilization, by IVF or artificial insemination (AI) [11]. However, there exists a great deal of variability in terms of the efficiencies reported in the literature, the cause of which has not been clear.

Transfection Reagent-Mediated Foreign DNA Transfer into Sperm

The method of directly incubating foreign DNA with sperm is simple, but frequently of low efficiency. Squires demonstrated a new method according to the principle of cell transformation in 1993. He packaged transgenes with liposomes then co-cultured this preparation with sperm *in vitro*. Positively charged liposomes and negatively charged transgenes form a liposome-DNA complex, with an enhanced ability to fuse with the sperm membrane. Subsequently, a number of experiments demonstrated that the use of liposomes as a package can significantly increase the efficiency of gene transfer. Both directly incubating transgenes with sperm and liposome-mediated transfer into sperm were comparted by Wang. The results demonstrated that the presence of liposome improved the transfection efficiency of transgene into sperm [19].

Besides liposomes, there are many other transfection reagents based on positive charge, such as activated dendrimers. Human coagulation factor VIII gene transgenic mice were successfully and efficiently produced by a SMGT technique incorporating activated dendrimers, and human coagulation factor VIII protein was expressed in the animals' bodies. The results showed that 3/9 F0 and 2/8 F1 progeny contained human coagulation factor VIII cDNA. Transcription and expression of human coagulation factor VIII cDNA occurred in the liver and kidneys of all the transgenic mice [20]. Purified human coagulation factor VIII protein from transgenic animals suggests new prospective strategy for treatment for hemophilia A.

Electroporation-Mediated Transgenes Transfer into Sperm

The application of an electric pulse(s) to a cell suspension induces polarisation of the membrane components of living cells and develops a voltage potential across the membrane. When the potential difference between the inside and the outside of the cell membrane passes a critical level, the membrane components are reorganized into pores in localized areas, and therefore the cells become permeable to the entry of macromolecules including transgenes [21-23]. This process of modifying the permeability of the cell membrane by an electric field is called electroporation, and it has been applied to sperm, as an augmented form of SMGT. The process of electroporation itself does not affect *in vitro* embryonic development. However, the number of oocytes fertilised with electroporated DNA-treated spermatozoa, and developing to the 16-cell stage is significantly lower compared to controls without DNA[15]. Several experiments from various groups have demonstrated that electroporation can increase the efficiency of DNA transfer into sperm [24-26]. However, electroporation can cause a premature acrosome reaction, and reduced sperm fertilisation ability [26].

DMSO-Sperm Mediated Gene Transfer

Shen *et al.* employed 3% final concentration of dimethylsulfoxide (DMSO) as a medium to transfect testicular germ cells with exogenous DNA *via* repeated direct injection into animal testis [27]. The new method called DMSO-sperm mediated gene transfer represents a high efficient and simple transgenic technology [28]. The results showed that the transgenic ratio using DMSO-sperm mediated gene transfer is 56.3% (by PCR and

Southern blot analysis). Compared with controls, the transgenic ratios were 39.6% and 47.8% using liposome-mediated SMGT employing Tfx™-50 Reagent or Lipefectamin™-2000, respectively. These results suggested that DMSO-SMGT is a powerful tool for mass generation of transgenic mice.

Virus-Sperm Mediated Gene Transfer

There are two commonly employed viral vectors for gene transfer, namely adenoviral and retroviral. With the exception of recent lentivirus-based retroviral vectors, most retroviral vectors developed for gene transfer applications can infect only dividing cells; by contrast, adenoviral vectors can infect both dividing and nondividing cells. Adenoviral vectors have relatively high transduction efficiency in target cells compared to retroviral vectors [29]. Adenoviral vectors can be prepared at higher titer than retrovirus vectors, and infect a large range of host cells, including hematopoietic stem cells and embryonic stem cells [30, 31]. Thus, adenoviral vectors have appeared as a promising choice for viral vector-based SMGT. However, there has been no evidence of infection after direct *in vitro* exposure of spermatogenic cells or mature sperm to adenoviral vectors [32]. This suggests that male germ cells, including spermatogonial stem cells, cannot be infected by adenovirus vectors.

The first reported evidence of germ-line transduction in males used retrovirus vectors, which have relatively high infection efficiency and have been widely used in the transduction of stem cells in several self-renewing tissues [33]. Spermatogonial stem cells were infected with retroviral vectors *in vitro* and transplanted into the seminiferous tubules for offspring production. Transplanted stem cells colonised the empty seminiferous tubules of infertile recipient testes and reinitiated spermatogenesis, eventually leading to the production of transgenic animals [34, 35]. This retroviral vector gene transfer method is efficient, easy to operate, and has high transgene integration rates; however the virus can trigger exogenous gene expression, the virus particle has a limited carrying ability, and retroviruses have various safety concerns.

Intracytoplasmic Sperm Injection (ICSI) Transgenesis

Intracytoplasmic Sperm Injection (ICSI) means micro-injecting a single sperm into the egg cytoplasm. ICSI can be used to create transgenic mice where the sperm has been pre-exposed to transgene molecules. This method can allow the sperm to carry so-called giant DNA molecules, such as yeast artificial chromosome (YAC) transgenes. In 2004, Moreira *et al.* obtained founders exhibiting germline transmission of an intact and functional transgene of 250 kilobases, by coinjecting spermatozoa and YACs into metaphase II oocytes [36]. When compared with the standard PNI method, ICSI is more efficient, and can express large transgenes. However, the original methods using freeze-thawed spermatozoa showed severe chromosomal damage and low offspring rates after embryo transfer. Recently, an improved method to generate transgenic mice efficiently has been described using a simple pretreatment of spermatozoa with 10 mM NaOH. These spermatozoa lost their plasma membrane and tail, while still maintaining nuclear integrity. Sperm heads were mixed with 0.5-5 ng/µl of the transgene for enhanced green fluorescent protein (EGFP) for 3 min to 1 h at room temperature and were then microinjected into oocytes by ICSI. The best results were obtained when treated spermatozoa were incubated with 2 ng/µl of EGFP for 10 min; 55.6% of injected embryos developed to the blastocyst stage, and more than half (56.9%) of them displayed EGFP fluorescence. Thus, a simple sperm pretreatment with NaOH before ICSI resulted in an efficient insertion of an exogenous gene into the host genome. This method allows for easy production of transgenic mice, requiring fewer oocytes for micromanipulation than classical methods [37].

TESTIS-MEDIATED GENE TRANSFER (TMGT)

Gene transfer *via* IVF of oocytes by spermatozoa that have been incubated with DNA-containing medium is the typical method of SMGT. It is simple, but the low efficiency leads to frequent negative results. In this situation, *in vivo* gene transfer through direct introduction of transgenes into testes is developing, because it is more straightforward (no need for IVF, just mate naturally), and more efficient (transgenes have been detected not only in F0, but in at least the second generation) [38]. Recently *in vivo* gene transfer to the whole tissue has become popular for the purpose biological studies and the future possibility of gene therapy, and, several strategies have been developed [39-41]. For the testis, transgene delivery mediated by adenoviral vector [42], or non-viral vectors such as liposomes [43] and electroporation [44, 45] have been applied to the mouse [46].

An *in vivo* approach to utilising spermatozoa as vectors, by transfering transgenes to the progeny *via* the testicular spermatozoa, a process termed testis mediated gene transfer (TMGT), has been employed by a number of groups. By this approach, DNA is injected directly into the testis with the expectation that testicular spermatozoa at various developmental steps may be more susceptible to DNA transfection than fully matured epididymal spermatozoa [43]. Chang *et al.* reported that liposome-encapsulated DNA injected into the testis could be transferred into the egg *via* sperm at fertilisation, and expressed in and transmitted to the descendants [47]. He also showed that, in his group's method of TMGT, exogenous DNA injected with liposomes is not integrated into the genome of the sperm and that the integration occurs after fertilisation [48]. Thus, the validity of sperm as vectors for TMGT may depend on a method suitable for loading spermatozoa with transgenes. Yonezawa first examined the efficiency of each of several different liposomes in associating transgenes with epididymal spermatozoa [49]. Then, to further establish optimum conditions for TMGT and to find a reliable application of this method, he investigated the expression of the introduced gene in early embryos, as well as the existence of the DNA in fetuses and pups at different ages. For some investigations, enhanced green fluorescent protein (EGFP) gene was used as a reporter to insert in a strong expression vector [50]. In 2002, scientists found that injection of EGFP expression vector/liposome complex into the testis resulted in transfection of epithelial cells of epididymal ducts facing the lumen, although the transfection efficiency appeared to be low. *In vivo* electroporation towards the caput epididymis immediately after injection of EGFP expression vector into a testis greatly improved the uptake of transgenes by the epididymal epithelial cells [51].

The mechanism of TMGT was found using testis injection with trypan blue (TB), Hoechst 33342 and plasmid DNA: when exogenous DNA is introduced directly into testis, it is rapidly transferred to epididymis *via* the rete testis and efferent ducts and then incorporated by epididymal spermatozoa, which will subsequently transfer the DNA to oocytes through fertilisation. Sato had developed a complete method of TMGT [38, 51]. For example, 70 μl of TB (Trypan Blue Stain 0.4%) was slowly injected with a 30-guage needle attached to a 1ml plastic disposable syringe at a depth of 5-6 mm through the capsule of the testis After injection, the needle was slowly removed (Fig. **2**). As with the mature sperm vector mentioned above, DMSO and electroporation can improve the efficiency of TMGT for the production of transgenic animals.

Figure 2: Schematic Representation of DNA Injection into Mouse Testis.

Injection of 70 μl of a solution was performed at the corner of the testis near the caput epididymis to a depth of 5-6 mm [51].

In vivo electroporation of foreign DNA injected into seminiferous tubules has become an efficient and convenient assay system for spermatogenic-specific gene expression during mouse spermatogenesis. Methodological modifications have been made to enhance the transfection efficiency and to generate transgenic offspring using GFP as a marker. Yamazaki has made a great contribution to this area [45]. His method is as follows: Male mice were anesthetized using pentobarbital sodium (75 mg/kg) injected intraperitoneally and the testis was exposed under the dissecting microscope. 0.4% Trypan-blue was added to the DNA solution to monitor the accuracy of the injection [38], and the DNA solution was injected into seminiferous tubules using an injection glass pipette, mostly as described previously [45]. To increase the injection volume, the injection pipette was connected to a polyethylene tube filled with the DNA solution, and injection was made until almost all the surface tubules (80–100%) became stained with Trypan-blue. After the DNA injection, electroporation was performed with an electrosquare porator T820 (BTX, San Diego, CA). Testis was held between a tweezers-type electrode and square electric pulses were applied eight times at 30–50 V with a time constant of 50 msec according to Muramatsu *et al.* [44].

CONCLUDING REMARKS

Over the past twenty years, the technology of murine SMGT has undergone rapid development, progressing from *in vitro* to *in vivo* approaches, with concomitant advantages of low expense and high convenience. However, the low transgene integration efficiency limits the application of this technology. If we overcome this disadvantage, SMGT will become very attractive in the future.

REFERENCES

[1]   Gordon JW, Scangos GA, Plotkin DJ, *et al.* Genetic transformation of mouse embryos by microinjection of purified DNA. Proc Nat Acad Sci USA 1980; 77: 7380-84.

[2]   Colman A. Somatic cell nuclear transfer in mammals: progress and applications. Cloning 1999-00;1:185-200.

[3]   Chan AWS. Transgenic animals: Current and alternative strategies. Cloning 1999; 1: 25-46.

[4]   Wolf E, Schernthaner W, Zakhartchenko V, *et al.* Transgenic technology in farm animals-progress and perspectives. Exp Physiol 2000; 85: 615-25.

[5]   Renard JP, Zhou Q, LeBourhis D, *et al.* Nuclear transfer technologies: Between successes and doubts. Theriogenology 2002; 57: 203-22.

[6]   Wu Z, Li Z, Yang J. Transient Transgene Transmission to Piglets by Intrauterine Insemination of Spermatozoa Incubated With DNA Fragments. Mol Reprod Dev 2008; 75: 26-32.

[7]   Wall RJ. New gene transfer methods. Theriogenology 2002; 57: 189-01.

[8]   Lavitrano M, Camaioni A, Fazio VM, *et al.* Sperm cells as vectors for introducing foreign DNA into eggs: genetic transformation of mice. Cell 1989; 57: 717-23.

[9]   Brinster RL, Sandgren EP, Behringer RR, *et al.* No simple solution for making transgenic mice. Cell 1989; 59: 239-41.

[10]   Chang K, Qin J, Jiang M, *et al.* Effective generation of transgenic pigs and mice by linker based spermmediated gene transfer. BMC Biotechnol 2002; 2: 5.

[11]   Lavitrano M, Busnelli M, Cerrito MG, *et al.* Sperm-mediated gene transfer. Reprod Fertil Dev 2006; 18: 19-23.

[12]   Lavitrano M, Forni M, Bacci ML, *et al.* Sperm mediated gene transfer in pig: Selection of donor boars and optimization of DNA uptake. Mol Reprod Dev 2003; 64: 284-91.

[13]   Yang J, Ratovitsk T, Brady JP, *et al.* Expression of myostatin pro domain results in muscular transgenic mice. Mol Reprod Dev 2001; 60: 351-61.

[14]   Yang J, Zhao B. Postnatal expression of myostatin propeptide cDNA maintained high muscle growth and normal adipose tissue mass in transgenic mice fed a high-fat diet. Mol Reprod Dev 2006; 73: 462-69.

[15]   Rieth A, Pothier F, Sirard MA. Electroporation of bovine spermatozoa to carry DNA containing highly repetitive sequences into oocytes and detection of homologous recombination events. Mol Reprod Dev 2000; 57: 338-45.

[16]   Brackett BG, Baranska W, Sawichi W, *et al.* Uptake of heterologous genome by mammalian spermatozoa and its transfer to ova through fertilization. Proc Nat Acad Sci USA 1971; 68: 353-57.

[17]   Spadafora C. Sperm cells and foreign DNA: a controversial relation. Bioessays 1998; 20: 955-64.

[18]   Maione B, Lavitrano M, Spadafora C, *et al.* Sperm-mediated gene transfer in mice. Mol Reprod Dev 1998; 50: 406-09.

[19]   Wang XM, Ruan Y, Xie QD, *et al.* Research on Gene Transfer by Spermatozoa. Yi Chuan 2005; 27: 195-200. [Article in Chinese]

[20]   Yin J, Zhang JJ, Shi GG, *et al.* Sperm mediated human coagulation factor VIII gene transfer and expression in transgenic mice. Swiss Med Wkly 2009; 139: 364-72.

[21] Knight DE, Scrutton MC. Gaining access to the cytosol: the technique and some application of electropermeabilization. Biochem J 1986; 234: 497-506.

[22] Tsong TY. Voltage modulation of membrane permeability and energy utilization in cells. Biosci Rep 1983; 3: 487-05.

[23] Serpeusu EH, Kinosita K Jr, Tsong TY. Reversible and irreversible modification of erythrocyte membrane permeability by electric field. Biochim Biophys Acta 1985; 812: 779-85.

[24] Horan R, Powell R, Bird JM, et al. Effects of electropermeabilization on the association of foreign DNA with pig sperm. Arch Androl 1992; 28: 105-14.

[25] Sin FY, Walker SP, Symonds JE, et al. Electroporation of salmon sperm for gene transfer: efficiency, reliability, and fate of transgene. Mol Reprod Dev 2000; 56: 285-88.

[26] Gagné MB, Pothier F, Sirard MA. Electroporation of bovine spermatozoa to carry foreign DNA in oocytes. Mol Reprod Dev 1991; 29: 6-15.

[27] Shen W, Li L, Pan Q, et al. Efficient and simple production of transgenic mice and rabbits using the new DMSO-sperm mediated exogenous DNA transfer method. Mol Reprod Dev 2006; 73: 589-94.

[28] Li L, Shen W, Min L, et al. Human lactoferrin transgenic rabbits produced efficiently using dimethylsulfoxidesperm-mediated gene transfer. Reprod Fertil Dev 2006; 18: 689-95.

[29] Takehashi M, Kanatsu-Shinohara M, Inoue K, et al. Adenovirus-mediated gene delivery into mouse spermatogonial stem cells. Proc Natl Acad Sci USA 2007; 104: 2596-601.

[30] Shui JW, Tan TH. Germline transmission and efficient DNA recombination in mouse embryonic stem cells mediated by adenoviral-Cre transduction. Genesis 2004; 39: 217-23.

[31] Neering SJ, Hardy SF, Minamoto D, et al. Transduction of primitive human hematopoietic cells with recombinant adenovirus vectors. Blood 1996; 88: 1147-55.

[32] Hall SJ, Bar-Chama N, Ta S, et al. Direct exposure of mouse spermatogenic cells to high doses of adenovirus gene therapy vector does not result in germ cell transduction. Hum Gene Ther 2000; 11: 1705-12.

[33] Nagano M, Shinohara T, Avarbock MR, et al. Retrovirus-mediated gene delivery into male germ line stem cells. FEBS Lett 2000; 475: 7-10.

[34] Nagano M, Brinster CJ, Orwig KE, et al. Transgenic mice produced by retroviral transduction of male germ-line stem cells. Proc Natl Acad Sci USA 2001; 98: 13090-95.

[35] Hamra FK, Gatlin J, Chapman KM, et al. Production of transgenic rats by lentiviral transduction of male germ-line stem cells. Proc Natl Acad Sci USA 2002; 99: 14931-36.

[36] Moreira PN, Giraldo P, Cozar P, et al. Efficient generation of transgenic mice with intact yeast artificial chromosomes by intracytoplasmic sperm injection. Biol Reprod 2004; 71: 1943-47.

[37] Li C, Mizutani E, Ono T, et al. An efficient method for generating transgenic mice using NaOH-treated spermatozoa. Biol Reprod 2010; 82: 331-40.

[38] Sato M, Yabuki K, Watanabe T, et al. Testis-mediated gene transfer (TMGT) in mice: Successful transmission of introduced DNA from F0 to F2 generations. Transgenics 1999; 3: 11-22.

[39] Marshall E. Gene therapy's growing pains. Science 1995; 269: 1050-55.

[40] Dzau VJ, Morishita R, Gibbons GH. Gene therapy for cardiovascular disease. Trends Biotechnol 1993; 11: 205-10.

[41] Thierry AR, Lunardi-Iskandar Y, Bryant JL, et al. Systemic gene therapy : biodistribution and long-term expression of a transgene in mice. Proc Natl Acad Sci USA 1995; 92: 9742-46.

[42] Blanchard KT, Boekelheide K. Adenovirus-mediated gene transfer to rat testis *in vivo*. Biol Reprod 1997; 56: 495-00.

[43] Kim JH, Jung-Ha HS, Lee HT, et al. Development of a positive method for male stem cell-mediated gene transfer in mouse and pig. Mol Reprod Dev 1997; 46: 515-26.

[44] Muramatsu T, Shibata O, Ryoki S, et al. Foreign gene expression in the mouse testis by localized *in vivo* gene transfer. Biochem Biophys Res Commun 1997; 233: 45-49.

[45] Yamazaki Y, Fujimoto H, Ando H, et al. *In vivo* gene transfer to mouse spermatogenic cells by deoxyribonucleic acid injection into seminiferous tubules and subsequent electroporation. Biol Reprod 1998; 59: 1439-44.

[46] Yamazaki Y, Yagi T, Ozaki T, et al. *In vivo* gene transfer to mouse spermatogenic cells using green fluorescent protein as a marker. J Exp Zool 2000; 286: 212-18.

[47] Chang K, Ikeda A, Hayashi K,et al. Production of transgenic rats and mice by the testis-mediated gene transfer. J Reprod Dev 1999; 45: 29-36.

[48] Chang K, Ikeda A, Hayashi K, et al. Possible mechanisms for the testis-mediated gene transfer as a new method for producing transgenic animals. J Reprod Dev 1999; 45: 37-42.

[49] Yonezawa T, Furuhata Y, Hirabayashi K, et al. Detection of transgene in progeny at different developmental stages following testis-mediated gene transfer. Mol Reprod Dev 2001; 60: 196-01.

[50]   Okabe M, Ikawa M, Kominami K, *et al.* 'Green mice' as a source of ubiquitous green cells. FEBS Lett 1997; 407: 313-19.

[51]   Sato M, Ishikawa A, Kimura M. Direct injection of foreign DNA into mouse testis as a possible *in vivo* gene transfer system *via* epididymal spermatozoa. Mol Reprod Dev 2002; 61: 49-56.

CHAPTER 7

Sperm-Mediated Gene Transfer in Aquatic Species: Present, Past and Future

Carlos Frederico Ceccon Lanes[1] and Luis Fernando Marins[2,*]

[1]*University of Nordland, Norway and* [2]*Universidade Federal do Rio Grande (FURG), Brazil*

Abstract: Transgenic research has been developed for several purposes including genetic improvement of aquacultured species and production of experimental models for biomedical research. Microinjection of DNA into pronuclei/nuclei of fertilised eggs is the technique that has been most used and applied successfully in transgenic animal production. However, this technique is laborious, time-consuming and it is limited by egg characteristics of some species. In aquatic animals such as marine finfish and shellfish the obstacles are even higher since these organisms have, in general, small and fragile eggs associated with high mortality during the first stages of development. Thus, en masse transgenesis methods such as Sperm Mediated Gene Transfer (SMGT) or Testis Mediated Gene Transfer (TMGT) could be useful for the production of genetically modified aquatic species. Although SMTG has been proved to be a good alternative for en masse transgenesis, this technique will only be routinely applicable in the near future if further investigations are focused on the factors that influence the interaction between spermatozoa and exogenous DNA as well as the processes that regulate transgene integration and expression.

Keywords: Aquaculture, DNA uptake, Expression, Finfish, Integration, Shellfish, Spermatozoa, Sperm mediated gene transfer, Testis mediated gene transfer, Transgenesis, Transgenic.

INTRODUCTION

The first transgenic aquatic animal was produced in 1985, when Zhu and colleagues microinjected the human GH gene into newly fertilised eggs of goldfish [1]. Since that time, gene transfer studies have been carried out in more than 35 teleost species and several aquatic invertebrates including sea urchins, oysters, clams, mussels, abalones and shrimps [2, 3]. Major areas of transgenic research in fish include transfer of genes coding for several proteins including: growth hormone to increase growth performance and feed conversion efficiency; antifreeze proteins for enhanced cold tolerance and freeze resistance; and antimicrobial peptides for increased disease resistance. Additional research goals have included transfer of metabolic genes to promote low-cost diets, and genetic methods for inducing sterility. Moreover, genetically modified fish have also been developed as experimental models for biomedical research, especially in studies involving embryogenesis and organogenesis, as well as in the study of human diseases, xenotransplantation and recombinant protein production for producing important therapeutic agents [3, 4].

By contrast, in aquatic invertebrates, such research is only in its initial stages due to complications with the introduction and expression of foreign genes. Work to date in this field has focused primarily on improving disease resistant and growth increase [3].

In fish, several techniques are currently available for gene transfer including: microinjection of DNA into the pronuclei/nuclei of fertilised eggs [5]; eggs and sperm electroporation [6, 7]; gonad lipofection [7]; retroviral vectors [8]; and particle gun bombardment [9]. Of these, microinjection is the technique that has been most used and applied successfully in transgenic fish production. However, microinjection is a laborious, time-consuming technique, and it is limited by the egg characteristics of some species (small and fragile eggs, hard chorion, little transparency, pronuclei invisible, *etc.*) [10]. Moreover, marine finfish and shellfish larvae have a high "natural" mortality during larviculture, representing another obstacle in the use

---

*Address correspondence to Luis Fernando Marins: Federal University of Rio Grande – FURG, Institute of Biological Sciences, Molecular Biology Laboratory, Av. Itália, Km 8, CEP 96203-900, Rio Grande, RS – Brazil; Tel: 55 53 3293-5191; E-mail: dqmluf@furg.br

of microinjection for gene transfer with marine animals. Therefore, sperm mediated gene transfer (SMGT) could be useful to achieve transgenesis in aquatic species with high fertility, small and fragile eggs, easy access to spermatozoa and huge production of sperm, such as finfish and shellfish.

Interest in SMGT for aquatic animals has increased due to several advantages related to other methods, such as: (*i*) huge numbers of oocytes can be fertilised with treated sperm; (*ii*) it overcomes some of the disadvantages of conventional gene transfer systems resulting from egg characteristics such as opaqueness, stickiness, buoyancy, invisible pronuclei and a tough chorion; (*iii*) fish sperm are easy to collect and handle, and simply adding water is enough to activate them; (*iv*) sperm of aquatic animals can be kept by cryopreservation so that treated sperm are always ready for usage; (*v*) SMGT is an inexpensive, rapid and simple method and does not require any particular equipment or ability and can be performed in field conditions [11, 12]. However, SMGT has not been routinely applied for producing of transgenic aquatic animals due to difficulties of transmission, integration and expression of foreign DNA by the embryo. Moreover, in aquatic species there is a lack of understanding regarding processes regulating the interaction between spermatozoa and exogenous DNA, hence many questions still remain without answers. In this chapter, we will review the advances and limitations of this technology in aquatic species.

MECHANISM OF BINDING OF SPERM TO DNA

The first results with SMGT in aquatic species were reported in 1989, when Arezzo [13] demonstrated that exogenous DNA could be transmitted to offspring by simply incubating sperm cells from sea urchin with transgenes prior to fertilization. In the same year, Lavitrano *et al.* [1989] reported similar results with mice. From the first moment, this new methodology was marred by difficulties of reproducibility, which eventually caused a good deal of skepticism over SMGT [14]. Subsequently, a number of studies have been carried out in order to elucidate the mechanisms that regulate the interaction between spermatozoa and foreign DNA. However, this work has been conducted mainly in the context of mammals. Unfortunately, therefore, in aquatic species the mechanisms involved in this process are still poorly understood.

In mammals, the mechanisms of DNA binding and internalisation by spermatozoa are highly regulated and specific. The main binding site of foreign DNA is in the sperm head in the subacrosomal region and in the proximity of the equatorial area. The exogenous DNA interacts with DNA-binding proteins (DBPs) of 30-35 KDa, which are present on the sperm cell surface [15, 16]. This high degree of affinity of sperm for DNA appears to be mediated by the complex structure of the MHC class II molecules and the antigen CD4, located in the posterior region of the mouse sperm head [17]. Once in the nucleus, the DNA becomes tightly associated to nuclear proteins and can be cleaved by sperm endonucleases, and finally integrated into the genome [18-20]. However, spermatozoa are naturally protected against the intrusion of foreign DNA molecules. At least three mechanisms have been identified that antagonise the binding of foreign DNA with spermatozoa: (*i*) an inhibitory factor 1 (IF-1), present in the seminal fluid of mammals or bound to the spermatozoa membrane, that prevents the binding of exogenous DNA [16]; (*ii*) a calcium-dependent DNase, also found in the seminal plasma of mammals and, according to Carballada & Esponda [21], being the main factor inhibiting DNA uptake by mammalian spermatozoa; and (*iii*) a sperm endogenous nuclease activity that is triggered in a dose-dependent manner upon interaction and/or internalisation of the exogenous DNA [19]. Therefore, these protections should minimise any unintentional interaction between sperm and exogenous sequences, which could compromise the sperm integrity and the genetic identity of the future progeny [22]. These mechanisms are discussed more fully in Chapter 12 of this book.

In aquatic species, at least IF-1 and calcium-dependent DNase should prevent the interaction between exogenous DNA and sperm cells. Zani *et al.* [16] have extracted IF-1 from sea urchin sperm and verified that this factor specifically interact with the DBPs, which prevents the binding of the exogenous DNA to spermatozoa. Recently, our group has demonstrated that seminal plasma of the Brazilian flounder (*Paralichthys orbignyanus*) contains a strong DNase activity [23]. In this work, DNase was present in all sources of semen tested: ejaculated semen, semen from the lumen and semen from macerated testes. Moreover, it was demonstrated that DNase activity occurs in a wide range of temperature (8-56°C), and acts very rapidly. However, DNase activity was decreased or eliminated by washing the spermatozoa with solutions containing EDTA. Even though sperm washing has been routinely used in mammals to remove seminal plasma in SMGT protocols, in most cases this protocol has not been applied for aquatic species.

Thus, IF-1 and DNase activity present in seminal plasma could be possible causes of failure in attempting to produce transgenic aquatic animals incubating directly exogenous DNA with spermatozoa.

TECHNICAL ASPECTS OF SMGT

In the last 20 years, many advances were carried out in SMGT and several important technical parameters affecting exogenous DNA uptake by spermatozoa were revealed. Semen quality, time and temperature of semen incubation with DNA, amount of exogenous DNA per sperm cell and DNA type (circular or linear) and architecture are parameters that should be seriously considered.

In 2003, Lavitrano and colleagues [24] demonstrated, in pigs, that the selection of appropriate sperm donors is an essential step for SMGT due to variation in sperm quality among the donors. It was verified that the DNA uptake is correlated with semen quality, particularly in terms of high progressive motility. Motility should be at least 80% initially and not less that 65% after washing procedures. In aquatic species, no detailed study has been carried out to verify the relation between the uptake of exogenous DNA by spermatozoa and semen quality. However, two studies show evidence that this relation should occur in aquatic animals. Firstly, Sin *et al.* [25] have observed variation in efficiency of gene transfer by electroporation from one reproductive season to another in salmon sperm. Secondly, Kuznetsov *et al.* [26] have verified that some mussels produce sperm, which have more capability of binding and transporting foreign DNA. Thus, factors that influence the sperm quality in aquatic species such as age of broodstock, hormonal induction of spermiation, sperm handling, broodstock nutrition, sperm contamination, stress and time in the reproductive season should be taken into consideration, in terms of their potential influence on SMGT success.

Another important factor is the time required by spermatozoa for binding and internalisation of exogenous DNA into their nuclei. Mostly in aquatic organisms, sperm cells have been incubated during around 30-60 min with exogenous DNA either for vertebrate (Table **1**) or invertebrate (Table **2**) species. In sea urchin, Arezzo [13] used ^{32}P-labeled nuclear DNA to verify the DNA incorporation into sperm cells. The results of this work indicated that within 75-90 minutes the ^{32}P incorporation by spermatozoa reached a maximum. In the same way, Guerra *et al.* [27], using rhodamine-labeled DNA, have demonstrated that most of mussel sperm cells were transfected (70%) within 2 hours. After short time periods (20-30 min) only 35-41% of the gametes appeared labeled, while longer times (4-6 h) did not increase the percentage of labeled gametes. Therefore, the ideal time of incubation to obtain the highest number of transgenic spermatozoa is around 90 min. Moreover, in aquatic species it is also necessary to establish how much DNA is internalised into sperm nuclei. According to Lavitrano *et al.* [24] approximately 20% of the sperm-bound DNA is internalised into sperm nuclei in pigs. In aquatic species, this information could help to improve the SMGT protocols, as well as elucidate in part the exogenous DNA integration and expression events, which are sometimes rare or nonexistent.

Table 1: Sperm Mediated Gene Transfer (SMGT) in Fish

Species	Technique	Time of Incubation	Examined Stage	% Transgenic	Integration	Expression	Reference
Zebrafish	DNA incubation	30-40 min	Adult	10.5-53.3	Yes	No (CAT)	[33]
Rainbow trout	DNA incubation	60-90 min	Larva	0	No	No (CAT)	[45]
	DNA incubation + 5% DMSO	60-90 min	Larva	0	No	No (CAT)	
African catfish	DNA incubation	40-60 min	Larva	0	No	n.d.[b]	[38]
	DNA incubation + electroporation + 10% DMSO	40-60 min	Larva	4.2	Yes	Yes (kanamycin)	
Tilapia	DNA incubation + electroporation	40-60 min	Larva	3.2	Yes	n.d.	[38]
Common carp	DNA incubation	40-60 min	Larva	0	No	n.d.	[38]
	DNA incubation + electroporation	40-60 min	Larva	2.6	Yes	n.d.	

Table 1: cont…..

Salmon	Electroporation	n.a.	Larva	5-10	No	n.d.	[70]
Salmon	Electroporation	n.a.	Larva	39	n.d.	n.d.	[34]
Loach	Electroporation	n.a.	Larva	50	Yes	Yes (GH)	[71]
Zebrafish	DNA incubation	30 min	Larva	6.5	Yes	n.d.	[41]
	DNA incubation + electroporation	30 min	Larva	14.5	Yes	n.d.	
Common carp	Shock osmotic + electroporation	n.a.	Juvenile	66	Yes	No (CAT)	[43]
Gilthead seabream	DNA incubation 10% DMSO	60 min	Sperm	n.d.	No	n.a.	[46]
	10% DMSO	60 min	Sperm	n.d.	Yes	n.a.	
	10% DMSO + freezing (-75°C)	n.a.	Sperm	n.d.	Yes	n.a.	
	10% DMSO + immersion in liquid nitrogen	n.a	Larva	n.d.	No	n.d.	
	Sonication Polyethylene glycol	n.a.	Larva	n.d.	No	n.d.	
		60 min	Sperm	n.d.	Yes	n.a.	
Rohu carp	Electroporation	n.a.	Larva	25	Yes	n.d.	[72]
Catla carp	Electroporation	n.a.	Larva	23	Yes	n.d.	[72]
Mrigal carp	Electroporation	n.a.	Larva	13	Yes	n.d.	[72]
Silver sea bream	Electroporation	n.a.	Juvenile	45	Yes	Yes (GH)	[7]
	TMGT[c] + lipofection	48 hours	Juvenile	68	Yes	Yes (GH)	
Grass carp	DNA incubation	10-30 min	Larvae	2.2-4.4	Yes	n.d.	[42]
	Electroporation	n.a.	Larvae	19.6-46.8	Yes	n.d.	
Rare minnow	Electroporation	n.a.	Larva	25.5-66.7	Yes	n.d.	[73]
Loach	Electroporation	n.a.	Larvae	5.7	n.d.	Yes (lacZ)	[74]
Rohu	Electroporation	n.a.	Juvenile	25	Yes	Yes (GH)	[75]
Brazilian flounder	DNA incubation	50 min	Sperm	n.d.	Yes	n.a.	[23]

[a]n.a., not applicable.

[b]n.d., not determined.

[c]TMGT, testis-mediated gene transfer.

The process of interaction between spermatozoa and exogenous DNA is also affected by the temperature of incubation, which influences sperm membrane permeability, DNA uptake, sperm viability and DNase activity. In general, incubation has been carried out at 18-20°C or in ice when SMGT protocols are applied for aquatic species. Incubation at low temperatures should be more appropriate, so that sperm quality is better conserved and DNase activity is reduced. However, spermatozoa permeability should be species-specific due to differences in the phospholipids and fatty acids composition of sperm among aquatic species [28]. Thus, the ideal temperature of incubation should be investigated for each species.

The amount of exogenous DNA per sperm cell is another limiting factor in transgenic animal production through SMGT. Although high concentrations of exogenous DNA increases the possibility of DNA uptake by spermatozoa and the number of transgenic spermatozoa, it also decreases sperm cell motility, viability and fertilising capacity [12, 29, 30]. In SMGT protocols published for aquatic species, it is difficult to carry out a comparison among the reports due to different number of spermatozoa used in each protocol, different units used to express the DNA quantity and on occasion the number of spermatozoa not being mentioned. In Brazilian flounder we have tested three different DNA concentrations (50, 100 and 500 ng) per 10^6

spermatozoa [23]. We verified a reduction of sperm motility when a concentration of 100 ng DNA/10^6 cells was used. Moreover, in the highest concentration (500 ng DNA/10^6 cells), spermatozoa motility was not observed, indicating that spermatozoa could be undergoing apoptosis as described for mice by Maione *et al.* [19]. According to these authors, high DNA concentrations (100-500 ng/10^6 cells) induce strong nuclease activity in sperm cells, yielding not only cleavage of the foreign DNA, but also degradation of sperm chromosomal DNA. Thus, a process resembling apoptosis is triggered in spermatozoa when increased concentrations of exogenous DNA are used for SMGT. Only the lowest DNA concentration (50 ng/10^6 cells) did not affect spermatozoa motility. Sperm motility is one of the primary measures of semen quality in fish and it is directly related to sperm fertilising capacity [31]. Therefore, the amount of DNA used for SMGT in fish should not exceed 50 ng/10^6 cells.

Regarding gene transfer by microinjection, it has been verified that linear DNA is 15 fold more efficient in transgenic fish production than circular DNA [32]. In a different way, when SMGT is applied the DNA type does not seem to affect the efficiency of uptake by sperm cells. Khoo *et al.* [33] and Symonds *et al.* [34] have evaluated the two forms of DNA (circular versus linear) and verified that DNA uptake by spermatozoa was not different. However, most studies performed in aquatic species have used linearised DNA. In mammals, two studies have demonstrated that the primary structure of plasmid and DNA architecture interferes in the SMGT efficiency. Firstly, Sciamanna *et al.* [35] have verified that plasmids of retrotransposon (pVLCNhGH) or viral (pCMVhGH) origin, affect differently the embryo development of pig and mice. In this study, the plasmid of retrotransposon origin caused massive embryo lethality, but increased the yield of genetic transformation among born animals compared to viral constructs. Secondly, Hoelker *et al.* [36] have demonstrated that the CMV-INF-τ-IRES-EGFP-nts plasmid, bearing an additional murine nontranscribed spacer (nts) insert, was 2-8 fold more efficient in incorporation into spermatozoa than same plasmid without nts. Moreover, the plasmid bearing nts increased IFN-τ mRNA levels in bovine blastocyts. In aquatic species, the influence of exogenous DNA origin and architecture in the SMGT efficiency have not been investigated yet and should be evaluated in the future.

MODIFICATIONS OF SPERM-MEDIATED GENE TRANSFER

The efficiency of SMGT by incubating spermatozoa directly with exogenous DNA for transgenic fish production has been low or nonexistent, either for vertebrate (Table **1**) or invertebrate (Table **2**) species. The only exception was reported by Khoo *et al.* [33], where high rates of transgenic zebrafish were obtained using sperm-incubation. Therefore, alternative techniques have been used to increase the uptake and binding of foreign DNA by sperm cells, as well as to increase the number of transgenic spermatozoa in aquatic animals. Electroporation, liposomes, dimethyl sulfoxide (DMSO) and injection of foreign DNA into the gonads of male, known as testis-mediated gene transfer (TMGT), are techniques currently in progress for application in aquatic species.

Table 2: Sperm Mediated Gene Transfer (SMGT) in Aquatic Invertebrates.

Species	Technique	Time of Incubation	Examined Stage	% Transgenic	Integration	Expression	Reference
Sea urchin	DNA incubation	1 hour	Embryo	n.d.[a]	n.d.	Yes (CAT)	[13]
Abalone	Electroporation	n.a.[b]	Larva	65	Yes	Yes (CAT)	[40]
Prawn	SMI[c]	2-6 hours	Larva	70	Yes	No (GFP)	[52]
Oyster	DNA incubation + electroporation	30 min	Larva	5.6-50	Yes	n.d.	[76]
Mussel	Electroporation	n.a.[b]	Larva	n.d.	Yes	n.d.	[26]
	1 % DMSO	30-60 min	Larva	0	No	n.d.	
	1% DMSO + heat shock (42°C, 1 min)	30-60 min	Larva	0	No	n.d.	
	1% DMSO + osmotic shock	30-60 min	Larva	0	No	n.d.	
	DNA incubation	30-60 min	Larva	n.d.	Yes	n.d.	

Table 2: cont....

Mussel	DNA incubation	2 hours	Sperm	65-70	Yes	n.a.	[27]
Claim	DNA incubation	2 hours	Sperm	58.5	Yes	n.a.	[27]
Abalone	TMGT[d]	3-6 hours	Adults	60	Yes	Increased growth	[53]
Shrimp	SMI + electroporation	n.d.	Adults	30	Yes	n.d.	[54]

[a]n.d., not determined.

[b]n.a., not applicable.

[c]SMI, spermatophore-microinjection.

[d]TMGT, testis-mediated gene transfer.

Electroporation has been the most-applied SMGT augmentation technique. This method utilises a series of short electrical pulses to induce formation of short-lived pores in the phospholipid bi-layer of the cell membrane, thereby permitting the entry of DNA molecules into the cells; once the electrical pulses are diminished, the pores on the cell membrane are re-sealed rapidly [37]. In 1992, Müller and colleagues [38] reported for the first time a successful application of electroporation of exogenous DNA into spermatozoa in fish. Since then, electroporation has been applied in different fish species and the efficiency of gene transfer has been found to be between 20% to 30% in most cases (Table **1**). However, an exception has been reported by Walker *et al.* [39], where optimal electroporation conditions rendered more than 90% of the newly fertilised salmon transgenic. In 1997, Tsai and colleagues [40] demonstrated the high potential of this technique for aquatic invertebrates such as abalone. In this study, Southern blot analysis showed the existence of exogenous DNA in 65% of trochophore-stage larva examined. In some cases, electroporation has reduced sperm motility and viability as well as the fertilising ability of the spermatozoa as demonstrated for salmon, zebrafish and rare minnow [34, 41, 42]. Therefore, electroporation parameters such as field strength, pulse length, pulse number, and ionic strength of the buffer, which affect sperm motility and uptake of DNA, should be optimised in a species-specific manner [34].

Another alternative to improve SMGT efficiency was tested by Kang *et al.* [43]. They examined the efficiency of an osmotic differential to promote the influx of extracellular medium and hence the uptake of foreign DNA into common carp sperm cells during electroporation. Firstly, spermatozoa were dehydrated in a hyperosmotic solution and subsequently rehydrated with a hyposmotic solution containing exogenous DNA, and during this process electroporation was applied. The success rate of gene transfer was 66% in 30-day-old fish analyzed through PCR.

Cationic lipids and DMSO have also been used to increase the DNA uptake by sperm cells. DMSO is routinely used for sperm cryopreservation in aquatic species. For marine fish, DMSO has been considered the best cryoprotectant due to its fast penetration into spermatozoa, as well as by interaction with the phospholipids of the membrane sperm avoiding cryoinjuries during the cryopreservation process [44]. In SMGT, DMSO has been used during the DNA incubation with spermatozoa in vertebrate (Table **1**) or invertebrate (Table **2**) species. However, all attempts failed to produce transgenic aquatic animals using this method, even though exogenous DNA was incorporated by spermatozoa [26, 45, 46]. Regarding cationic lipids, this method has been frequently used for *in vitro* as well as *in vivo* gene transfer due to the ability of selected lipids to interact with DNA [47, 48]. The lipid/DNA complexes are formed due to interaction between positively charged lipid molecules and the negatively charged DNA molecules, producing liposomes or lipoplexes. Liposome-mediated gene transfer *via* sperm cells has been successfully carried out in rabbit, chicken, equine and bovine animals [36, 49-51]. In aquatic species, the incubation of exogenous DNA with spermatozoa in the presence of cationic lipids has not been tested yet. However, Lu *et al.* [7] have injected a liposome-DNA mixture into testes of sea bream at least 48 hours before spawning. The injected males were crossed with active females and the rate of gene transfer ranged from 59% to 76% in hatched fry, as verified through PCR analysis. Moreover, the efficiency of gene transfer was improved more than 80% by injecting multiple doses of the liposome-transgene mixture.

A promising approach for transgenic animal production is direct introduction of foreign DNA into male gonads. However, fish testes are not easily accessible and require minor surgery to be exposed. Conversely, aquatic invertebrates such as mussels and shrimps provide simple access for gonad manipulation. Li & Tsai [52] have demonstrated that exogenous DNA can be transferred by microinjection into extruded spermatophores of giant freshwater prawn. For *in vitro* fertilisation, the treated spermatophore was placed into the female spermatheca. Southern blot and PCR analyses revealed the existence of foreign DNA in 70% of larvae analyzed. Chen *et al.* [53] also used the same procedure to transfer foreign DNA into the Japanese abalone testis. In this study, the efficiency of gene transfer was 90%, 92.5% and 60% for larvae, juveniles and 1-year-old adults, respectively. Moreover, around 20% of mature G_0 animals contained the transgene in their gonad. Recently, Chen *et al.* [54] have demonstrated that the sperm inside the spermatophore can be electroporated with foreign DNA. In this procedure, the intact mature spermatophore was extracted from American white shrimp males, injected with DNA and electroporated. Sperm from the electroporated spermatophore were squeezed out and transferred to the thylecum of gravid females. Foreign DNA was found in 30% of 8-month-old shrimps analyzed through PCR. Consequently, Testis Mediated Gene Transfer (TMGT) has come to be considered a simple, minimally invasive, high-throughput, non-viral based, non-sperm damaging, efficient approach for producing transgenic aquatic invertebrates [52, 53].

In 2006, Anzar & Buhr [30] demonstrated that the transfection efficiency of bull spermatozoa was higher in frozen-thawed than in fresh spermatozoa of same ejaculate. According to them, cryopreservation alters the plasma membrane of spermatozoa, facilitating the binding and internalisation of exogenous DNA, leading to a higher transfection rate. Thus, DNA incubation after sperm cryopreservation could be another good alternative to increase the DNA uptake for aquatic species, since at least 200 fish species and 30 aquatic invertebrate species have had their sperm successfully cryopreserved [44, 55].

INTEGRATION, EXPRESSION AND TRANSMISSION OF FOREIGN DNA

In aquatic species, all published work reports DNA incorporation by spermatozoa and, in some cases, nuclear internalisation of foreign DNA [41]. Moreover, efficient alternatives to increase the DNA uptake by spermatozoa have been provided. However, integration, transmission and mainly expression of foreign DNA by embryo have been rare and, until now, the events that occur after fertilisation with transgenic sperm have been poorly understood.

In fish, some experiments showed that exogenous DNA can be eliminated in the course of embryo development. Liu *et al.* [46] failed to detect (by PCR or dot blot) clear and consistent evidence for sperm-mediated transfer of foreign DNA to hatched gilthead seabream larvae. Foreign DNA was evident sporadically up until 16 hours post-fertilisation. At 40 hours post-fertilisation and after hatching no signal was detected in the individuals analyzed. The authors have suggested that the foreign DNA detected in earlier stages would be traces of exogenous DNA on the egg chorion remained during the fertilisation. Later, Khoo [56] studied the time course of DNA following transfer to zebrafish embryos. Dot blot results have shown that plasmid content peaks at 10 hours (late gastrula) after fertilisation, but decline on further development, indicating the occurrence of a protective mechanism that degrades exogenous DNA after gastrulation. In most fish species, the whole sperm head and the tail enter the ova during fertilisation [57], so that either externally bound or internal foreign DNA might be introduced into ova. Therefore, it is possible that a natural mechanism of defense against foreign DNA be activated after fertilization, resulting in the elimination or degradation of foreign DNA during the course of embryo development. This mechanism should be further investigated in aquatic species with external fertilisation.

There are few published works addressing the transmission of exogenous DNA beyond the F_0 generation using SMGT in aquatic species. In zebrafish, the exogenous DNA was transferred for F_1 and F_2, but no Mendelian ratios were observed. In addition, there was no evidence of integration and expression of exogenous gene, with the introduced DNA persisting extrachromosomaly [33, 56]. The occurrence of transgenic sequences persisting outside the chromosomes and transmitted to the next generation has been also reported in birds, mammals and insects [58-61]. However, the persistence and transmission to the progeny of extrachromosomal transgenic structures are not yet understood [22]. The integration and transmission of transgenes to progeny was successfully demonstrated by Lu *et al.* [7] in silver sea bream (*Sparus sarba*). Approximately 37% of progeny derived from founders produced by sperm electroporation and 27% of progeny derived from founders produced by TMGT were transgenic. Nevertheless, these results suggest that transgenic founders were germ-line mosaic.

Usually, when microinjection is used to produce transgenic fish it almost invariably produces mosaic fish due to delayed transgene integration, which occurs only after a few cycles of embryonic cell division. If the transgene is integrated in only one cell group or tissue but not into germ cells then its transmission to descendants is impossible [62]. Theoretically, sperm manipulation could diminish the mosaicism rate, facilitating the exogenous DNA integration into the genome of the spermatozoa or zygote. However, applications of SMGT methods for aquatic species have resulted in mosaicism as much as has microinjection. In the case of microinjection, many attempts have been reported to improve genomic integration of the transgene and, consequently, decrease the mosaicism rate. These include mixing of exogenous DNA with nuclear transport peptide [63], DNA flanked with inverted terminal repeats from adeno-associated viruses [64], and DNA introduced together with the *I-SceI* meganuclease [65]. All these methods could be alternatives to improve the integration and expression of exogenous DNA using SMGT for aquatic species.

Interestingly, in some cases the exogenous DNA has been detected through PCR or Southern blot analysis in later development stages, but gene expression has not occurred in the stages concerned [33, 52, 66]. According to Khoo [56], the lack of expression can be explained due to several reasons including rearrangement of the transgene sequence after being internalised into the sperm, lack of functionality of promoters used with reporter genes, and extrachromosomal presence of the transgene.

CONCLUDING REMARKS

During the last 20 years, a good deal of scientific effort has been devoted towards the goal of producing transgenic aquatic animals through of SMGT. However, most of these studies have focused only on the direct application of SMGT for transgenic production or on the implementation of different methods to increase the uptake of exogenous DNA by spermatozoa. Few studies have tried to evaluate the factors that influence the interaction between spermatozoa and exogenous DNA. In the opposite way, studies performed in mammals have been conducted with the goal of understanding the process and practical factors which affect the internalisation of exogenous DNA by spermatozoa. Consequently, for aquatic species a number of questions still remain to be answered regarding the SMGT use. Future studies will need to answer important scientific and technical questions so that this method of gene transfer may become a reliable technique for the production of transgenic aquatic animals. At least six major questions should be addressed in future studies, as follows: (*i*) Are MHC class II and CD4 complex responsible by the high degree of affinity between sperm and exogenous DNA in aquatic species (as they are in mammals)? (*ii*) How much exogenous DNA is incorporated by spermatozoa nuclei? (*iii*) How does sperm quality affects the interaction with DNA? (*iv*) What kind of mechanism protects the embryo after fertilisation? (*v*) Is the exogenous DNA remaining outside of spermatozoa responsible for activation of this mechanism? (*vi*) How is the extrachromosomal DNA transmitted to the next generation?

It is clear that foreign DNA can associate with and even penetrate sperm cells independently of the aquatic species used. However, this association can affect sperm motility and viability and, consequently, may decrease fertilisation success. Usually, in fish a high sperm to egg ratio is used during fertilisation. As a result, it is possible that during this stage the substantial loss of sperm viability may result in a population of spermatozoa in which surviving fertile sperm tend to be those with little or no uptake of foreign DNA. Therefore, an accurate and precise method, possibly based on existing computer-assisted sperm analysis (CASA) systems, should be used to choose the best donor and to evaluate sperm motility after DNA incorporation. The application of this technique will assist the improvement of SMGT protocols in fish.

In fish, electroporation is the most common technique used for increasing exogenous DNA uptake by spermatozoa. Nevertheless, we also believe that the use of cationic lipids and cryopreserved cells could be excellent alternatives or enhancements to improve the efficiency of exogenous DNA uptake by spermatozoa. In lipofection, the lipid/DNA complex is formed as micelle-like structure and this protect the exogenous DNA from DNase activity present in seminal plasma [67, 68]. Moreover, the use of cationic lipids simplifies the SMGT protocol since sperm washing is not necessary. According to Kang *et al.* [66], the washing step is the main factor that affects sperm quality and is thus one of the reasons for repeated failure in SMGT. Regarding cryopreserved cells, this method should be interesting for aquatic species in which the synchronisation of gamete availability of both sexes is difficult. Furthermore, cryopreservation allows the use of high quality sperm produced during the reproductive season. However, fundamental factors such as DNase activity and the interaction between spermatozoa and exogenous DNA after cell cryopreservation should also be actively investigated.

The TMGT technology described in this review seems to be the best alternative to the direct use of spermatozoa as vectors of exogenous DNA transfer in aquatic invertebrates. The use of this technique might solve the problems of introduction and expression of foreign genes due to the promising results obtained in shrimp, abalone and prawn [52-54]. In fish, although Lu *et al.* [7] have successfully produced transgenic silver sea bream, the protocol might need to be refined for each species used. Information on factors affecting success need to be quantified; such factors include the volume of exogenous DNA to be injected into the testis, the number of times that the testis should be manipulated before spermiation, the correct period of injection before spermiation, as well as the procedure to manipulate the testis without affecting animal survival. In general, fish spermatozoa have a highly condensed chromatin that is envisaged as being difficult for genomic integration of foreign DNA. However, TMGT applied in the early stages of spermatogenesis should promote contact between exogenous DNA and germ cells in different developmental stages. In theory, this could improve integration and expression of foreign DNA. According to Schulz *et al.* [69], data on the timing of spermatogenetic events are scarce, but some studies have verified that the duration of meiosis plus spermatogenesis varies from 7 to 21 days in tropical species and from 1 to 3 months in species living in temperate and cold zones. Therefore, testis manipulation should be started in fish at least 3 to 4 weeks before spermiation.

In summary, SMGT is a good alternative for transgenic marine species production since it offers the possibility of en masse gene transfer. In aquatic species, positive results have been obtained, but many questions remain to be answered. Therefore, this approach will only be routinely applicable in the near future if new investigations are conducted, focusing on factors that influence the interaction between spermatozoa and exogenous DNA as well as the processes that regulate transgene integration and expression during embryo development.

REFERENCES

[1]   Zhu Z, Li G, He L, Chen S. Novel gene transfer into the fertilised eggs of gold fish (Carassius auratus L. 1758). J Appl Ichthyol 1985; 1: 31-34.

[2]   Zbikowska HM. Fish can be first - advances in fish transgenesis for commercial applications. Transgenic Res 2003; 12: 379-89.

[3]   Rasmussen RS, Morrisey MT. Biotechnology in aquaculture: transgenics and polyploidy. Compr Rev Food Sci Food Saf 2007; 6: 2-16.

[4]   Maclean N. Genetically modified fish and their effects on food quality and human health and nutrition. Trends Food Sci Technol 2003; 14: 242-52.

[5]   Ozato K, Kondoh H, Inohara H, Iwamatsu T, Wakamatsu Y, Okada TS. Production of transgenic fish: introduction and expression of chicken δ-crystallin gene in medaka embryos. Cell Differ 1986; 19: 237-44.

[6]   Inoue K, Yamashita S, Hata J, *et al.* Electroporation as a new technique for producing transgenic fish. Cell Differ Dev 1990; 29: 123-28.

[7]   Lu JK, Fu BH, Wu JL, Chen TT. Production of transgenic silver sea bream (Sparus sarba) by different gene transfer methods. Mar Biotechnol 2002; 4: 328-37.

[8]   Lin S, Gaiano N, Culp P, Burns JC, Friedmann T, Yee JK. Integration and germ-line transmission of a pseudotyped retroviral vector in zebrafish. Science 1994; 265: 666-69.

[9]   Yazawa R, Hirono I, Yamamoto E, Aoki T. Gene transfer for Japanese flounder fertilised eggs by particle gun bombardment. Fish Sci 2005; 71: 869-74.

[10]   Hostetler HA, Peck SL, Muir WM. High efficiency production of germ-line transgenic Japanese medaka (Oryzias latipes) by electroporation with direct current-shifted radio frequency pulses. Transgenic Res 2003; 12: 413-24.

[11]   Lavitrano M, Camaioni A, Fazio VM, Dolci S, Farace MG, Spadafora C. 1989. Sperm cells as vectors for introducing foreign DNA into eggs - genetic transformation of mice. Cell 1989; 57: 717-23.

[12]   Tsai HJ. Electroporated sperm mediation of a gene transfer system for finfish and shellfish. Mol Reprod Dev 2000; 56: 281-84.

[13]   Arezzo F. Sea-urchin sperm as a vector of foreign genetic information. Cell Biol Int Rep 1989; 13: 391-04.

[14]   Brinster RL, Sandgren EP, Behringer RR, Palmiter RD. No simple solution for making transgenic mice. Cell 1989; 59: 239-41.

[15]   Lavitrano M, French D, Zani M, Frati L, Spadafora C. The interaction between exogenous DNA and sperm cells. Mol Reprod Dev 1992; 31: 161-69.

[16]   Zani M, Lavitrano M, French D, *et al.* The mechanism of binding of exogenous DNA to sperm cells - factors controlling the DNA uptake. Exp Cell Res 1995; 217: 57-64.

[17]   Lavitrano M, Maione B, Forte E, *et al.* The interaction of sperm cells with exogenous DNA: a role of CD4 and major histocompatibility complex class II molecules. Exp Cell Res 1997; 233: 56-62.

[18]   Magnano AR, Giordano R, Moscufo N, Bacetti B, Spadafora C. Sperm/DNA interaction: Integration of foreign DNA sequences in the mouse sperm genome. J Reprod Immunol 1998; 41: 187-96.

[19]   Maione B, Pittogi C, Achene L, Lorenzini R, Spadafora C. Activation of endogenous nucleases in mature sperm cells upon interaction with exogenous DNA. DNA and Cell Biol 1997; 16: 1087-97.

[20]   McCarthy S, Ward WS. Interaction of exogenous DNA with the nuclear matrix of live spermatozoa. Mol Reprod Dev 2000; 56: 235-37.

[21]   Carballada R, Esponda P. Regulation of foreign DNA uptake by mouse spermatozoa. Exp Cell Res 2001; 262: 104-13.

[22]   Smith K, Spadafora C. Sperm-mediated gene transfer: applications and implications. BioEssays 2005; 27: 551-62.

[23]   Lanes CFC, Sampaio LA, Marins LF. Evaluation of DNase activity in seminal plasma and uptake of exogenous DNA by spermatozoa of the Brazilian flounder Paralichthys orbignyanus. Theriogenology 2009; 71: 525-33.

[24]   Lavitrano M, Forni M, Bacci ML, *et al.* Sperm mediated gene transfer in pig: Selection of donor boars and optimisation of DNA uptake. Mol Reprod Dev 2003; 64: 284-91.

[25]   Sin FYT, Walker SP, Sin IL, Symonds JE, Mukherjee UK, Khoo JGI. Electroporation of salmon sperm for gene transfer: Efficiency, reliability, and fate of transgene. Mol Reprod Dev 2000; 56: 285-88.

[26]   Kuznetsov AV, Pirkova AV, Dvoryanchikov GA, *et al.* Study on the transfer of foreign genes into the mussel Mytilus galloprovincialis Lam. eggs by spermatozoa. Russ J Dev Biol 2001; 32: 254-62.

[27]   Guerra R, Carballada R, Esponda P. Transfection of spermatozoa in bivalve molluscs using naked DNA. Cell Biol Int 2005; 29: 159-64.

[28]   Drokin SI. Phospholipids and fatty-acids of phospholipids of sperm from several freshwater and marine species of fish. Comp Biochem Physiol Biochem Mol Biol 1993; 104: 247-62.

[29]   Sasaki S, Kojima Y, Kubota H, Tatsura H, Hayashi Y, Kohri K. Effects of the gene transfer into sperm mediated by liposomes on sperm motility and fertilisation *in vitro*. Hinyokika Kiyo 2000; 29: 6-15.

[30]   Anzar M, Buhr MM. Spontaneous uptake of exogenous DNA by bull spermatozoa. Theriogenology 2006; 65: 683-90.

[31]   Rurangwa E, Kime DE, Ollevier F, Nash JP. The measurement of sperm motility and factors affecting sperm quality in cultured fish. Aquaculture 2004; 234: 1-28.

[32]   Penman DJ, Beeching AJ, Penn S, Maclean N. Factors affecting survival and integration following microinjection of novel DNA into rainbow trout eggs. Aquaculture 1990; 85: 35-50.

[33]   Khoo HW, Ang LH, Lim HB, Wong KY. Sperm cells as vectors for introducing foreign DNA into zebrafish. Aquaculture 1992; 10: 1-19.

[34]   Symonds JE, Walker SP, Sin FYT, Sin IL. Development of a mass gene transfer method in chinook salmon: optimisation of gene transfer by electroporated sperm. Mol Mar Biol Biotechnol 1994; 3: 104-11.

[35]   Sciamanna I, Piccoli S, Barberi L, *et al.* DNA dose and sequence dependence in sperm-mediated gene transfer. Mol Reprod Dev 2000; 56: 301-05.

[36] Hoelker M, Mekchay S, Schneider H, *et al.* Quantification of DNA binding, uptake, transmission and expression in bovine sperm mediated gene transfer by RT-PCR: Effect of transfection reagent and DNA architecture. Theriogenology 2007; 67: 1097-07.

[37] Tieleman DP. The molecular basis of electroporation. BMC Biochem 2004; 5: 10.

[38] Müller F, Ivics Z, Erdélyi F, *et al.* Introducing foreign genes into fish eggs with electroporated sperm as a carrier. Mol Mar Biol Biotechnol 1992; 1: 276-81.

[39] Walker SP, Symonds JE, Sin IL, Sin FYT. Gene transfer by electroporated chinook salmon sperm. J Mar Biotechnol 1995; 3: 232-34.

[40] Tsai HJ, Lai CH, Yang HS. Sperm as carrier to introduce an exogenous DNA fragment into the oocyte of Japanese abalone (Haliotis divorsicolor suportexta). Transgenic Res 1997; 6: 85-95.

[41] Patil JG, Khoo HW. Nuclear internalisation of foreign DNA by zebrafish spermatozoa and its enhancement by electroporation. J Exp Zool 1996; 274: 121-29.

[42] Zhong JY, Wang YP, Zhu ZY. Introduction of the human lactoferrin gene into grass carp (Ctenopharyngodon idellus) to increase resistance against GCH virus. Aquaculture 2002; 214: 93-01.

[43] Kang JH, Yoshisaki G, Homma O, Strüssmann CA, Takashima F. Effect of an osmotic differential on the efficiency of gene transfer by electroporation of fish spermatozoa. Aquaculture 1999; 173: 297-07.

[44] Suquet M, Dreanno C, Fauvel C, Cosson J, Billard R. Cryopreservation of sperm in marine fish. Aquac Res 2000; 31: 231-43.

[45] Chourrout D, Perrot E. No transgenic rainbow trout produced with sperm incubated with sperm linear DNA. Mol Mar Biol Biotechnol 1992; 1: 282-85.

[46] Liu XY, Zohar Y, Knibb W. Association of foreign DNA with sperm of gilthead seabream, Sparus aurata, after sonification, freezing and dimethyl sulfoxide treatments. Mar Biotechnol 1999; 1: 175-83.

[47] Ross PC, Hensen ML, Supabphol R, Huis W. Multilamellar cationic liposomes are efficient vectors for *in vitro* gene transfer in serum. J Liposome Res 1998; 8: 499-20.

[48] Vitiello L, Bockhold K, Joshi PB, Worton RG. Transfection of cultured myoblasts in high serum concentration with DODAC:DOPE liposomes. Gene Ther 1998; 5: 1306-13.

[49] Wang HJ, Lin AX, Chen YF. Association of rabbit sperm cells with exogenous DNA. Anim Biotechnol 2003; 14: 155-65.

[50] Yang CC, Chang HS, Lin CJ, *et al.* Cock spermatozoa serve as the gene vector for generation of transgenic chicken (Gallus gallus). Asian-australas J Anim Sci 2004; 17: 885-91.

[51] Ball BA, Sabeur K, Allen WR. Uptake of exogenous DNA by equine spermatozoa and applications in sperm-mediated gene transfer. Anim Reprod Sci 2006; 94: 115-16.

[52] Li SS, Tsai HS. Transfer of foreign gene to giant freshwater prawn (Macrobrachium rosenbergii) by spermatophore-microinjection. Mol Reprod Dev 2000; 56: 149-54.

[53] Chen HL, Yang HS, Huang R, Tsai HJ. Transfer of a foreign gene to Japanese abalone (Haliotis diversicolor supertexta) by direct testis-injection. Aquaculture 2006; 253: 249-58.

[54] Chen TT, Chen MJ, Chiou T, Lu JK. Transfer of foreign DNA into aquatic animals by electroporation. In: Nakamura H, Eds. Electroporation and Sonoporation in Developmental Biology. Springer, Tokyo, Japan, 2009; pp. 229-37.

[55] Gwo JC. Cryopreservation of aquatic invertebrate semen: A review. Aquac Res 2000; 33: 259-71.

[56] Khoo HW. Sperm-mediated gene transfer studies on zebrafish in Singapure. Mol Reprod Dev 2000; 56: 278-80.

[57] Ginzberg AS. Fertilisation in fishes and the problem of polyspermy. Translated from Russian by Blake Z. and Golek B. Jerusalem: Israel Program for Scientific Translations 1972; p. 366.

[58] Rottman OJ, Antes R, Hofer P, Maierhofer G. Liposome mediated gene transfer *via* spermatozoa into avian egg cells. J Anim Breed Genet 1992; 109: 64-70.

[59] Robison KO, Ferguson HJ, Cobey S, Vaessin H, Smith BH. Sperm-mediated transformation of the honey bee, Apis mellifera. Insect Mol Biol 2000; 9: 625-34.

[60] Kuznetsov AV, Kuznetsova IV, Schit L. DNA interaction with rabbit sperm cells and its transfer into ova *in vitro* and *in vivo*. Mol Reprod Dev 2000; 56: 292-97.

[61] Celebi C, Auvray P, Benvegnu T, Plusquellec D, Jegou B, Guillaudeux T. Transient transmission of a transgene in mouse offspring following *in vivo* transfection of male germ cells. Mol Reprod Dev 2002; 62: 477-82.

[62] Maclean N. Regulation and exploitation of transgenes in fish. Mutat Res 1998; 399: 255-66.

[63] Liang MR, Aleström P, Collas P. Glowing zebrafish: integration, transmission, and expression of a single luciferase transgene promoted by noncovalent DNA-nuclear transport peptide complex. Mol Reprod Dev 2000; 55: 8-13.

[64]   Hsiao CD, Hsieh FJ, Tsai HJ. Enhanced expression and stable transmission of transgenes flanked by inverted terminal repeats from adeno-associated virus in zebrafish. Dev Dyn 2001; 220: 323-36.

[65]   Thermes V, Grabher C, Ristoratore F, *et al.* I-SceI meganuclease mediates highly efficient transgenesis in fish. Mech Dev 2002; 118: 91-98.

[66]   Kang JH, Hakimov H, Ruis A, Friendship RM, Buhr M, Golovan SP. The negative effects of exogenous DNA binding on porcine spermatozoa are caused by removal of seminal fluid. Theriogenology 2008; 70: 1288-96.

[67]   Murphy EA, Waring AJ, Haynes SM, Longmuir KJ. Compaction of DNA in an anionic micelle environment followed by assembly into phosphatidylcholine liposomes. Nucleic Acids Res 2000; 28: 2986-92.

[68]   Sato F, Soh T, Hattori M, Fujihara N. Evaluation of deoxyribonuclease activity in seminal plasma of ejaculated chicken semen. Asian J Androl 2003; 5: 213-16.

[69]   Schulz RW, de França LR, Lareyre JJ, *et al.* Spermatogenesis in fish. Gen Comp Endocrinol 2010; 165: 390-11.

[70]   Sin FYT, Bartley AL, Walker SP, *et al.* Gene-transfer in chinook salmon (Oncorhynchus tshawytscha) by electroporating sperm in the presence of pRSV-LacZ DNA. Aquaculture 1993; 117: 57-69.

[71]   Tsai HJ, Tseng FS, Liao IC. Electroporation of sperm to introduce foreign DNA into the genome of loach (Misgurnus anguillicaudatus). Can J Fish Aquat Sci 1995; 52: 776-87.

[72]   Sarangi N, Mandall AB, Bandyopadhyay AK, Venugopal T, Mathavan S, Pandian TJ. Electroporated sperm-mediated gene transfer in Indian major carps. Asia Pac J Mol Biol Biotechnol 1999; 7: 151-58.

[73]   Zhong JY, Mao WF, Zhu ZY. Introduction of foreign DNA into the genome of rare minnow by sperm electroporation. Yi Chuan Xue Bao 2002; 29: 128-32.

[74]   Andreeva LE, Sleptsova LA, Grigorenko AP, Gavrushkin AV, Kuznetsov AV. Loach sperm cells transfer foreign DNA expressed in early development. Russ J Genet 2003; 39: 627-29.

[75]   Venugopal T, Anathy V, Kirankumar S, Pandian TJ. Growth enhancement and food conversion efficiency of transgenic fish Labeo rohita. J Exp Zool A Comp Exp Biol 2004; 301A: 477-90.

[76]   Hu W, Yu DH, Wang YP, Wu KC, Zhu ZY. Electroporation of sperm to introduce foreign DNA into the genome of Pinctada maxima (Jameson). Sheng Wu Gong Cheng Xue Bao 2000; 16: 165-68.

CHAPTER 8

Sperm-Mediated Gene Transfer in Agricultural Species

Joaquin Gadea[1,*], Francisco Alberto Garcia-Vazquez[1], Sebastian Canovas[1] and John Parrington[2]

[1]*University of Murcia, Spain and* [2]*University of Oxford, United Kingdom*

Abstract: The importance of transgenic farm animals resides in their usefulness for very different objectives: for instance in human medicine (to obtain pharmaceutical products, organs for xeno-transplantation, or as models for research in gene therapy) or for agricultural applications such as improved output of the carcass or milk production and composition, increased growth rate, improved feed utilisation, enhanced reproductive performance, increased prolificacy, as well as to enhance disease resistance or reduce environmental impact. Pronuclear DNA microinjection has long been the most reliable method to produce transgenic animals. However, although transgenic animals have been generated using this approach, it has many limitations. Sperm Mediated Gene Transfer (SMGT) is based on the ability of sperm to bind, internalise, and transport exogenous DNA into an oocyte during fertilisation. In this chapter we review the state of art of SMGT in farm animals with a special emphasis on porcine and bovine animals, with additional information related to other ruminants and horses. We evaluate the possible applications of transgenic pigs and cattle and review the factors related to the success of SGMT in these species and offer our own experience based on studies analyzing the main factors in porcine and bovine SMGT.

Keywords: Transgenic Pig, Transgenic Cattle, Biomolecules production, Human diseases models, Xenotransplantation, Artificial insemination, *In vitro* fertilization, ICSI, Farm animals, Recombinase, Embryo transfer, Flow cytometry, DNA binding.

INTRODUCTION

The term "transgenic animals" is used to denote animals with altered characteristics resulting from direct changes to their genetic material. Genes can be manipulated artificially, so that certain characteristics of an animal are changed, using different genetic techniques such as gene knockout (in which genes are made inoperative), knockin (in which genes are modified) or knockdown (in which expression of genes is reduced). The importance of transgenic farm animals resides in their usefulness for very different objectives: for instance in human medicine (to obtain pharmaceutical products, organs for xeno-transplantation, or as models for research in gene therapy) or for agricultural applications such as improved output of the carcass or milk production and composition, increased growth rate, improved feed utilisation, enhanced reproductive performance, increased prolificacy, as well as to enhance disease resistance or reduce environmental impact [1, 2]. The mouse has served as a model mammal in studies of human physiology or diseases, but its use as a model for humans has many limitations. For instance, factors such as lifespan, size, similarities in physiology and possibly genomic organisation make cattle and pigs more similar to humans than mice [3, 4] and they may provide a better model to study human diseases. In addition cows have the advantage that they produce a great amount of milk for a long period of time and it is relatively easy to recover secreted products of interest from their milk. In fact, the expression of proteins with potential therapeutic applications in the milk of livestock species appears to be one of the most attractive commercial applications of animal transgenesis [5]. As an example of this, recently some groups have reported the production of cloned transgenic cows with high levels expression of bioactive human growth hormone, human lactoferrin or human albumin in their milk [6-8].

Since the first reports about transgenic animals in the 1970s many applications still await solutions for technical or ethical challenges; however, other products of transgenic livestock technology will very soon

*Address correspondence to Joaquin Gadea: Department of Physiology, University of Murcia, Murcia 30.100, Spain; Tel: 34-868-88 46 55; E-mail: jgadea@um.es

be on the market [9]. The production of high-value pharmaceutical substances is the principal and most promising application for transgenic pigs, goats, sheep and cattle [10]. Some worldwide companies are investing great efforts in this promising technology. Due to its large-scale volume production, milk is the preferred vehicle for expression of proteins in transgenic animals, although they can be produced in a variety of biological fluids such as urine, saliva, blood and seminal fluid [11, 12].

Pronuclear DNA microinjection has long been the most reliable method to produce transgenic animals since 1985 when the first generation of transgenic farm animals were reported [13, 14]. However, although transgenic animals have been generated using this approach, it has many limitations: thus it is expensive and laborious, has a low success rate (1-4%) and leads to random integration of the DNA into the genome [5, 15]. Different methodologies have been developed to improve the efficiency of transgenesis over the last decade and a half. These include infection with retroviral vectors; nuclear transfer with modified somatic cells and sperm mediated gene transfer.

Transfection with retroviral vectors is one of the best choices for creation of transgenic animals because of their high gene transfer efficiency [16, 17], although they have some limitations (size limit for the transgene of ~10 kb and possibility of gene silencing). Transgenic animals have been successfully generated by the injection of virus into the periviteline space of oocytes (cattle, pigs, and monkeys), zygotes (cattle, mice, and rats), and early preimplantation embryos (cattle, mice, rat) [16, 18]. However, the health risk is the main barrier to the commercial use of this system. Another alternative for generating genetically altered animals is to first make the modification in somatic cells and then use the modified cells as donors for nuclear transfer (NT). Successful nuclear transfer (NT) of cultured cells, first demonstrated in cattle by [19], has provided an alternative for obtaining genetically modified animals. Unfortunately the procedure is inefficient and frequently leads to animals that are abnormal. The cause of these abnormalities is probably established during the first cell cycle after the NT.

Sperm Mediated Gene Transfer (SMGT) is based on the ability of sperm to bind, internalise, and transport exogenous DNA into an oocyte during fertilisation [20, 21]. Rabbit sperm cells were reported to spontaneously take up and transfer DNA into an oocyte during fertilisation resulting in the genetic modification of the two cell-stage embryos [20]. In 1989, the birth of live transgenic mice was reported after epidydimal sperm cells were used to transfer exogenous plasmid DNA into an oocyte during fertilisation [21]. SMGT has been used more or less successfully in the production of transgenic embryos and animals in a large number of species [22-26]. Although transgenic animals have been obtained using SMGT, the repeatability of this technique is low, with only a few numbers of laboratories reporting clearly positive results. In addition, inter-and intra-species success variability is still an unsolved problem associated with this technology. Also the mechanism whereby the foreign DNA is bound and internalised inside the sperm head, the mode of transmission upon fertilisation and the perseverance of the foreign DNA in the resulting embryos, fetuses and offspring, are issues that have not yet been fully elucidated-although the reader's attention is drawn to Chapter 12 where possible models are discussed.

Testis mediated gene transfer (TMGT) is a more recent system that introduces transgenes directly into the reproductive tract of male animals, either as naked DNA or encapsulated in liposomes. The sperm take up the transgenes *in vivo* (directly or through progression from sperm precursor cells) and transmit them *via* natural mating or artificial insemination [26-28]. Injection into the rete testis appears to be more suitable for *in vivo* gene transfer by electroporation than direct intratesticular injection [29].

In this chapter we review the state of art of SMGT in farm animals with a special interest in porcine and bovine animals, with additional information related to other ruminants and horses. We evaluate the possible applications of transgenic pigs and cattle and review the factors related to the success of the SGMT in these species and offer our own experience based on studies analysing the main factors in porcine and bovine SMGT.

TRANSGENIC CATTLE

In bovines a number of different methods have been used to obtain transgenic animals. The pronuclear injection approach was used for the first successful attempt to produce a transgenic bovine [30, 31];

however this technique has serious limitations as noted already. The success of pronuclear injection with respect to transgene integration is only 1% for cattle, pigs and sheep [2, 15] and around 1000 bovine zygotes must be microinjected to produce one founder transgenic animal [32]. In addition to the poor efficiency of pronuclear microinjection, this method usually results in a high percentage of mosaics in which not all cells of the animal contain the transgene [16]. The time and cost of screening for germline transmission in mosaic animals like cattle can prohibit generation of more transgenic animals through breeding. The success of pronuclear microinjection is evident in the generation of many transgenic cattle lines but its limitations have hindered the progress of bovine transgenics [33].

In the 1990s the success of approaches using mouse embryonic stem (ES) cells led to hopes that this might provide another route for generating transgenic cattle while circumventing many of the limitations of other transgenic techniques. Many attempts have been made to isolate ES cells in sheep or pigs with limited success [34-37]. In cattle success was achieved when ES-like cells were isolated from early embryos, transfected with exogenous DNA, reintroduced into preimplantation embryos and shown to contribute to tissues of the resulting calves [38]. However, genetic modification of cattle ES-like cells is not easy due to the culture system required to maintain them as undifferentiated ES cells and their inability to expand clonally *in vitro* [39]. Unfortunately the generation time and maintenance cost of multiple mosaic animals have prohibited testing for germline transmission using this approach.

Due to absence of proven ES cells and the recent advances in somatic cell nuclear transfer (SCNT) following the birth of Dolly the sheep, much current emphasis for creating bovine transgenics has been placed on this approach. One advantage of this technique over microinjection is that the sex of the animal can be predetermined by choosing the donor material (male or female) which would both increase efficiency and facilitate the manipulation of milk production through transgenesis [40]. Besides, the proper use of this technique would ensure that 100% of animals produced are transgenic and that every cell of an animal would have the transgene. However, the success rate for SCNT averages between 1-3% in most animals including cattle [41, 42], and the majority of embryos are lost during pregnancy with a 60% higher fetal loss between gestational days 35-60 when compared to embryos created through *in vitro* fertilisation (IVF) [43].

At first, the principle objective of studies producing transgenic cattle through SCNT was to alter milk composition. The first transgenic cattle produced through SCNT were transgenic for a fusion β-galactosidase-neomycin gene which demonstrated that cattle could be produced using transgenic cultured cells as donors for SCNT [39]. Subsequently [40] used SCNT to create calves transgenic for two casein genes involved in milk protein production.

Success in generating bovine embryos using transgenic cultured cells has differed between studies. While some studies have reported a decrease in blastocyst development rate using transgenic cells compared to nontransgenic cells for SCNT [44], others have not reported a difference in SCNT blastocyst development rate when using such transgenic cells [40, 45].

TRANSGENIC BOVINES GENERATED USING SPERM MEDIATED GENE TRANSFER (SMGT)

In bulls the binding of exogenous DNA to spermatozoa has been demonstrated by several groups [46-51] although the mechanism of binding and internalisation of exogenous DNA is an issue that has not been solved. Some studies have resulted in the creation of bovine transgenic embryos [50, 52-57] or calves [49, 55, 58] using SMGT, although the efficiency of this technique is quite low and huge differences between studies were observed (Table **1**).

Although the precise mechanics of binding is not clearly known, it is evident that the binding of exogenous DNA to spermatozoa is possible. However, the integration of the transgene into the genome is a major limiting factor using this approach. For these reasons, strategies such as electroporation of spermatozoa [52, 54] and the use of chemical transfection reagents like liposomes [55] or FuGene6 [56] have been developed to try and increase foreign DNA uptake into spermatozoa and integration into the genome.

Table 1: Bovine Sperm-DNA Interaction and Sperm-Mediated Gene Transfer Experiments

Year	Reference	Experimental	Characteristics
1990	Castro *et al.* [46]	DNA incubation	Radio labeling Formaldehyde inhibition
1991	Perez *et al.* [58]	AI	Transgenic embryos
1991	Atkinson *et al.* [47]	DNA incubation	Radio labeling Fresh and cold shocked spermatozoa
1991	Gagne *et al.* [52]	IVF	Electroporation Transgenic embryos
1992	Camanioni *et al.* [48]	DNA incubation	Frozen spermatozoa Heparine inhibition
1993	Francolini *et al.* [122]	DNA incubation	Radiolabeling
1995	Schellander *et al.*[49]	AI IVF	Radiolabeling Transgenic calf
1996	Rottman *et al.* [123]	AI	Liposome Transgenic calf
1996	Sperandio *et al.* [53]	IVF	Radiolabeling Transgenic embryos
2000	Chan *et al.* [124]	DNA incubation	Rhodamine labelled Confocal microscope
2000	Rieth *et al.* [54]	IVF	Electroporation
2000	Shemesh *et al.* [55]	IVF AI	REMI Transgenic calves
2003	Saritisatien *et al.* [59]	IVF ICSI	Epididimal and ejaculated
2006	Alderson *et al.* [50]	IVF	Protamines
2006	Anzar *et al.* [51]	DNA incubation	Fresh and frozen
2007	Hoelker *et al.* [56]	IVF	Liposomes, Fugene
2008	Pereyra-Bonnet *et al.* [57]	ICSI	Transgenic embryos
2009	Zi *et al.* [78]	DNA incubation IVF	Retrovirus
2009	Feitosa *et al.* [72]	DNA incubation	Frozen semen
2010	Canovas *et al.* [148]	DNA incubation	Frozen semen Flow cytometry

Gagne *et al.* [52] used electroporation to enhance introduction of DNA into bovine spermatozoa and obtained a fivefold increase in the percentage of plasmids bound. Rieth *et al.* [54] also combined electroporation with this approach to good effect, although the motility of sperm was significantly reduced after electroporation. When homologous recombinant offspring were studied the percentages of positive embryos were much higher (55%) than in the control group (9%). Other authors have showed that the use of liposomes is particularly effective in transferring DNA into bovine sperm. Shemesh *et al.* [55] produced transgenic bovine sperm using restriction enzyme mediated integration (REMI) to facilitate integration of exogenous DNA, in conjunction with lipofection. The results of IVF showed that 30% of the morula expressed the transgene (GFP) as determined by RT-PCR. When artificial insemination (AI) was used 50% of the cows became pregnant and 100% (2/2) of the calves tested demonstrated specific green fluorescence in 60% of their lymphocytes.

Hoelker *et al.* [56] showed that the ability of bull spermatozoa to bind exogenous DNA was not increased by treatment with liposomes, however the treatment with transfection reagent FuGene6 significantly increased (by 10-350 fold) binding of exogenous DNA. When the incorporation of exogenous plasmid into spermatozoa was analyzed it was found to significantly increase either following addition of liposome (15-27 fold) or FuGene 6

(50-120 fold). In addition, the cleavage rate and development up to blastocyst stage were not affected by either treatment; although the blastocysts derived contained 7-8 fold more plasmid when FuGene6 was used. According to these results the lipidic structure of the spermatic plasma membrane may influence binding of exogenous DNA. Indeed, some groups have proposed that there is a relationship between the binding of exogenous DNA to spermatozoa and the level of alteration of the plasma membrane [51, 59] or capacitation status [60]. It is known that capacitation involves changes in the distribution and composition of lipids, phospholipids (PL) and other molecules that provoke a destabilisation process and leads to an increase in the membrane fluidity and changes in its architecture, where cholesterol plays a key role [61]. Recently we have demonstrated [62] that capacitation using heparin increases the number of spermatozoa bound to DNA and also the proportion of live and DNA-bound spermatozoa relative to the control. We thus propose that capacitation using heparin increases the binding of exogenous DNA to bovine spermatozoa.

Different factors that affect the effectiveness of SMGT have been identified. It is known that seminal fluids of mammals block the binding of exogenous DNA [63] and epididymal spermatozoa bound in higher proportion to bovine spermatozoa [59]. With bovines it has been reported that frozen-thawed spermatozoa are more effective than fresh spermatozoa at taking up exogenous DNA [51], leading to the hypothesis that cryopreservation induces changes in the plasma membrane that facilitate the binding and internalisation of exogenous DNA.

We analyzed the dynamics of bovine sperm DNA binding and viability simultaneously using flow cytometry [64]. This has allowed us to determine the number of live spermatozoa carrying exogenous DNA that could fertilise oocytes in conventional IVF or AI systems. These studies showed that a proportion of live spermatozoa can also bind exogenous DNA (with around 15% of the sperm population possessing this capacity), in contrast with previous reports in pigs [65-68] which showed that most of the exogenous DNA was bound to dead or altered cells, a difference which could reflect different DNA binding patterns between species.

Previous studies suggested that sperm DNA binding could require an incubation time higher than 30 minutes [46, 48, 53]; however, our results showed that bovine spermatozoa were able to bind exogenous DNA much more rapidly, the binding process being complete in 30 minutes, These results suggest that long incubations with exogenous DNA before fertilisation are not necessary for maximal fertilisation results, particularly when considering that prolonged incubation negatively affects sperm viability.

We also observed that the pattern of binding was highly male-dependent. Significant differences between males were observed in the absolute capacity of sperm DNA binding, although the dynamics of binding followed a similar pattern. In contrast, previous authors [51] did not find that the donor of the spermatozoa affected spontaneous uptake of DNA, but they used fluorescent microscopy to evaluate a very limited numbers of cells. Our results are closer to agreement with results seen in earlier reports using pigs [25, 69] that showed that sperm cells from boars differed greatly in their capacity to take up exogenous DNA, although it was also observed that the dynamics were similar between males [70]. According to Celebi *et al.* [71] the condensation of DNA in spermatozoa shows intra-and interspecies variations, which may explain, in part, the differences in efficiency observed from one species to another.

Although nearly 100 papers have been published concerning SMGT during the last 20 years, the effect of exogenous DNA on sperm functionality remains unclear. Motility is a special characteristic of spermatozoa that is absolutely necessary for its biological function. A putative negative effect of DNA binding on motility could explain the failure of SMGT reported by numerous authors. Previous studies reported opposing results regarding DNA effect; one set of studies described a negative effect of the exogenous DNA on the motility of bovine spermatozoa [49, 51, 72], while others did not report any effect [50, 54]. Our recent results [73] demonstrated that incubation with exogenous DNA (5 µg/ml) decreases total and progressive sperm motility in bovines. Moreover, sperm motion parameters such as curvilinear, straight-line and average path velocities, linearity of the curvilinear trajectory, linearity, and beat cross-frequency were reduced after incubation with DNA.

In bovine studies, the efficiency of SMGT can vary widely depending on both the transgene and the gene transfer method [53, 54]. Recently Hoelker *et al.* [56] investigated whether the architecture of the foreign plasmid affects SMGT efficiency in bovines by comparing two plasmids. This group tested the effectiveness at each step by performing real time PCR to quantify the number of plasmids in sperm and embryos and the abundance of target gene transcripts in blastocysts after SMGT. Their results showed that exogenous plasmid binding to spermatozoa was significantly affected by the architecture of the plasmid (which contained the murine nontranscribed spacer fragment, nts). This architecture significantly increased binding and the incorporation of exogenous plasmid DNA to the sperm surface, the transfer of exogenous DNA to oocytes, and expression in bovine embryos at blastocyst stage.

One of the potential disadvantages of SMGT, which uses a "natural" vector, the sperm cell, for transporting exogenous DNA into the oocyte, is that nature provides mechanisms to protect spermatozoa against the intrusion of foreign DNA molecules, given that the reproductive tracts contain free DNA molecules (resulting from cell death and breakage) so that during physiological reproduction evolutionary chaos does not result. One barrier against intrusion of exogenous molecules is a sperm endogenous nuclease activity that is triggered in a dose-dependent manner upon interaction of spermatozoa with foreign molecules [74]. Apoptosis could also be a natural phenomenon to prevent transmission of exogenous DNA to the next generation.

In bovine SMGT, the degradation of the transgene by exogenous nucleases could be one limiting factor as has been observed in murine and porcine SMGT [74]. In addition, sperm tails have been shown to possess DNase II like activity [75]. Consequently protecting the transgene against nuclease activity might be expected to increase the efficiency of SMGT. Protamine sulfate has been shown to enhance lipid mediated transfection and protect against nuclease activity in mammalian cells [76]. Recently Alderson *et al.* [50] reported that after sperm transfection in bovines the plasmid remained intact, in contrast with swine spermatozoa where there is extensive degradation of the plasmid upon incubation with naked DNA for 1h [74]. However, Alderson *et al.* [50] showed that protamine sulfate did not improve the efficiency of bovine SMGT.

Other alternatives such as restriction enzyme-mediated integration (REMI) permit decondensation of the genomic DNA and facilitate the integration of the exogenous DNA into the genome. This enzyme and lipofection method has been applied successfully to cattle by [55] using IVF and artificial insemination. Overall, a number of problems remain to be solved, although these novel methods of active transgenesis could improve the effectiveness of bovine SMGT and lead to the development of new strategies.

Finally the use of ICSI in combination with the sperm mediated gene transfer methodology makes feasible the use of sperm with plasma membranes damaged by physical (freezing and thawing) or chemical methods (by using a detergent like Triton X-100). Disruption of the sperm membrane allows DNA constructs to associate with submembrane structures, and this is a key step for successful DNA insertion into oocytes [77]. The experience in bovines is limited to negative results using epididymal spermatozoa [59] and some recent preliminary results reported recently [57]. Also, new approaches in bovine SMGT could be pursued in the next years, such as the use of viral vectors to enhance DNA uptake, as has recently has been investigated in the yak [78].

PORCINE

In recent years, transgenic pigs have become an important tool in biomedical research, for instance for the production of bio-molecules in the mammary gland, the development of transgenic animals to improve productivity, for research into xenotransplantation, and as models for human diseases [79, 80] (Tables **2-4**). All these applications depend on the output of transgene(s) expression, and many different strategies such as pronuclear microinjection, vector virus and nuclear transfer have been developed to generate transgenic animals [81, 82] (Table **5**). DNA pronuclear microinjection has been the most popular system to generate transgenic pigs. However, besides being expensive, this technique is still inefficient when used to generate transgenic farm animals [22, 26].

Table 2: Some Important Milestones in the Application of the Technologies for Producing Transgenic Pigs

Methodology	Charasteristics	Reference
Pronuclear injection	First transgenic pigs generated by microinjection	Hammer *et al.* 1985; Brem *et al.* 1985 [13, 14]
Avian retrovirus	Porcine transgenic fetus generated by avian retrovirus	Petters *et al.* 1987 [125]
Sperm-Mediated Gene Transfer	Transgenic pigs generated by SMGT	Gandolfi *et al.* 1989 [83]
Pronuclear injection	Use of oocytes matured and fertilised *in vitro*	Kubish *et al.* 1995 [126]
Sperm-Mediated Gene Transfer	Transgenic pigs expressing DAF	Lavitrano *et al.* 1997 [85]
Moloney murine leukemia virus	Oocyte transfection	Cabot *et al.* 2001 [18]
Nuclear Transfer	First transgenic pigs generated by NT	Park *et al.* 2001 [127]
Sperm-Mediated Gene Transfer and ICSI	First transgenic pigs generated by SMGT-ICSI	Lai *et al.* 2001 [101]
Nuclear Transfer	First transgenic KO pigs	Lai *et al.* 2002 [128]
Equine infectious anaemia virus	Embryos injected in the peri-vitelline space	Whitelaw *et al.* 2004 [129]

Table 3: Approaches to Generating Transgenic Pig as Models for Human Diseases

Biomodels	Methodology	Reference
Retinitis pigmentosa	Pronuclear injection	Peters *et al.* 1997[130]
Cardiovascular: endothelial nitric oxide synthase	Nuclear Transfer	Hao *et al.* 2006 [131]
Cystic Fibrosis	Nuclear Transfer	Roger *et al.* 2008 [132]
Diabetes mellitus	SMGT-ICSI and Nuclear Transfer	Umeyama *et al.* 2009 [133]
Alzheimer	Nuclear Transfer	Kragh *et al.* 2009 [134]

Table 4: Approaches to Generating Transgenic Pig for Pharmaceutical Production in the Mammary Gland, Blood or Tissues.

Product	Methodology	Reference
Human Hemoglobin	Pronuclear injection	Swanson *et al.* 1992 [135]
Protein C	Pronuclear injection	Velander *et al.* 1992 [136]
Factor VII	Pronuclear injection	Paleyanda *et al.* 1997 [137]
Factor IX	Pronuclear injection	Van Cott *et al.* 1999 [138]
Human Albumin	SMGT-ICSI	Naruse *et al.* 2005 [107]
Human Erythropoietin	Pronuclear injection	Park *et al.* 2006 [139]
Factor von Willebrand	Pronuclear injection	Lee *et al.* 2009 [140]

Table 5: Approaches to Generating Transgenic Pigs for Agricultural Production

Objective	Characteristics	Methodology	Reference
Improve daily weight gain and feed efficiency	Bovine-GH	Pronuclear injection	Pursel *et al.* 1989 [141]
Resistance to influenza virus infection	Murine Mx-l protein	Pronuclear injection	Müller *et al.* 1992 [142]
Improve milk production in sows	Bovine α-Lactalbumin	Pronuclear injection	Bleck *et al.* 1998 [143]
Increase lean and reduce fat tissue	Insulin-like growth factor I	Pronuclear injection	Pursel *et al.* 1999 [144]
Reduce phosphorus manure	Expressing salivary phytase	Pronuclear injection	Golovan *et al.* 2001[145]
Induce to synthesize unsaturated fatty acids	Δ12 fatty acid desaturase	Pronuclear injection	Saeki *et al.* 2004 [146]
Increase omega-3 (n-3) fatty acids	n-3 fatty acid desaturase	Nuclear Transfer	Lai *et al.* 2006 [147]

The first investigation of SMGT in the pig was reported by Gandolfi *et al.* 1989 [83]. They incubated the sperm in presence of DNA and using them for artificial insemination they obtained animals that presented a positive signal after Southern blot analysis. However, the transgene was not transmitted to F1. Later, this group [84] reported an inability to produce transgenic pigs by this approach and intrauterine artificial insemination. In a separate study, Sperandio *et al.* (1996) [53] reported differences in the percentage of transgenic embryos obtained after artificial insemination according to the different constructs used, ranging from 0 to 5.7%. Using DNA they obtained 5 of 82 transgenic pigs (6.1%) detected by PCR, however no signal was found in sperm from one of the transgenic males. Later the same group reported the production of embryos and piglets that expressed the hDAF gene [85, 86] and stable transmission to progeny [87]. In the same year Sciamanna *et al.* (2000) reported the generation of hGH piglets with a high efficiency (66%) using a construct of retrotransposon origin [88]. Lavitrano *et al.* reported and confirmed high levels of transgenesis after artificial insemination [25, 70, 89] or laparoscopic insemination for the generation of mono or multi-transgenic pigs [90-92]. Also they reported the successful use of episomal vector [93]. However other groups have reported a transient transgene transmission to the offspring that later is not detected after 70-100 days, which could be the result of episomal DNA replications during the early stages of development [94].

Several factors could determine the successful of results using this technique including boar donors, construct used, incubation media, etc [68, 70, 95]. The efficiency of this technique depends on the kind of construct used [53, 70, 88] and the amount of DNA used. Thus, Sciamanna *et al.* 2000 reported that greater than 50-100 ng DNA/10^6 sperm (according to the kind of construct analysed) induced a lethal effect on the embryos and reduced the number of piglets born after artificial insemination [88]. It is known that the presence of exogenous DNA can induce toxic effects on the sperm functionality [74, 88], but under our experimental conditions it was observed that sperm incubation with foreign DNA did not involve an alteration of the seminal parameters [96]. These results are in accordance with another report which confirmed that porcine spermatozoa can be transfected with the exogenous DNA and maintain their motility [69].

DNA BINDING

Castro *et al.* [46] reported for the first time that foreign DNA can bind to porcine sperm after only 15 min of incubation and after 45 min each sperm cell is bound to 9 DNA molecules meaning that 0.2-0.3% of the total DNA present in the culture is bound to the sperm cells. Later, Horan *et al.* [97] extended these findings when they reported that some 380 DNA molecules remained bound to each sperm after a rigorous regime of washes. *In situ* hybridisation studies showed clearly that up to 30% of motile sperm carried DNA mainly on the post-acrosomal region of the sperm. The acrosome reaction induced by calcium ionophore did not modify the capacity of binding to DNA [98]. Gandolfi *et al.* reported lower values of binding, ranging from 12-17 % and they did not find any difference between different sperm treatments [84].

The presence of seminal plasma is known to inhibit the DNA binding; thus the elimination of the seminal plasma by washing the samples prior to the DNA incubation increased the binding to DNA [48, 70, 99]. The seminal plasma plays an important role acting as a natural barrier and protecting the spermatozoa from exogenous molecules that could compromise the integrity of the spermatozoa. So, the epididymal spermatozoa are a valuable model to explore the possible effect of the seminal plasma components. We have analysed the use of epididymal spermatozoa and obtained similar DNA-binding ability as with ejaculated spermatozoa [66].

Other studies have used 'augmentation' techniques, such as electroporation, antibodies, or liposomes, to 'force' sperm to capture transgenes [100-102]. It was demonstrated that this DNA binding percentage could be increased 10-15 fold if the sperm-DNA mixtures were subjected to a permeabilising pulse [100], and transgenic pigs have been generated after the use of a monoclonal antibody used as a cross linker to facilitate the binding of exogenous DNA to sperm. [102] Recently, the use of magnetic nanoparticles and liposomes have been reported as an efficiently way to introduce a transgene into embryos *via* spermatozoa [103].

On the other hand, determination of DNA binding to the sperm cells is a key point in this technique. Radiolabelling measurements [48, 70, 89], fluorescent microscopy [51], and immunohistochemistry [84] have been used to measure DNA binding. Flow cytometry seems to be a very valuable tool for evaluating

DNA binding and viability, so it is possible to evaluate the kinetics of the binding process [65, 67, 68]. This methodology has been validated with fluorescent microscopy observations and the use of multispectral imaging flow cytometry (ImageStream; Amnis Corporation, Seattle, WA), a combination of quantitative image analysis and flow cytometry.

Using flow cytometry, we have detected that most of the exogenous DNA is bound to dead cells or to cells with severe membrane alterations (PI stained), the percentage of DNA bound to fresh semen being close to 30% after 2 hours of coincubation. This is in concordance with other authors employing a non-radioactive method [69, 84], but lower than the data reported when radiollabelled measurements are carried out [48, 70, 89]. It has been previously reported that there is a window of opportunity in which the exogenous DNA binds to spermatozoa and it coincides with the early stage of capacitation [70]. It may be that under our experimental conditions, sperm capacitation occurs very quickly, so the DNA binding can occur but the acrosome reaction takes place too early and thus the spermatozoa die. In this case, changes in the sperm membrane functionality might modulate the time of DNA binding; the populations of viable sperm attached to the DNA may determine the success or failure in the production of transgenic animals in this way. On the other hand, it is possible that the DNA binding induces the alteration and death of the cell by endonuclease activation in an apoptotic-like process. The apoptosis of the spermatozoa could be a natural phenomenon to prevent the transmission of exogenous DNA to the following generation [51].

IVF ICSI

Some studies have demonstrated a lack of production of embryos expressing transgenes when they were generated by IVF [98] (Horan *et al.* 1992b: Lai *et al.* 2001); by contrast, recently some positive results have shown the presence of the exogenous DNA in embryos produced by IVF; however no expression of transgenes was detected [103, 104].

Another innovation in SMGT has been the use of intracytoplasmic sperm injection (ICSI) to deliver transgene-containing sperm cells directly into the egg, a process which was reported for the first time in mice [77] and later reported in rhesus monkeys and in pigs [65, 76, 101, 105-108].

In pigs, ICSI is a technique with potential applications in diverse fields of animal production and biomedicine. The combination of the ICSI-mediated method and *in vitro*-matured (IVM) oocytes would both greatly reduce the cost of ICSI-SMGT and streamline the procedure and would facilitate an expansion of the practical value of transgenic pigs, increasing their availability. Efficient ICSI-SMGT makes feasible the use of sperm with plasma membranes damaged by physical (freezing and thawing) or chemical methods (by using a detergent like Triton X-100). Disruption of the sperm membrane allows DNA constructs to associate with submembrane structures, and this is a key step for successful DNA insertion into oocytes [65, 77]

Lai *et al.* (2001) reported for the first time the production of embryos by ICSI that expressed EGFP (29.4 %) using sperm incubated with DNA and liposomes [101]. However they did not obtain any piglets after the embryo transfer of these embryos. Later Kurome *et al.* [106] reported the possibility of using IVM oocytes which would reduce the costs of oocytes/embryo collection from donor females. Naruse reported the use of this methodology to produce a transgenic pig expressing human albumin in the liver [107] and later the same group reported the combined use of ICSI-SMGT and with somatic cell nuclear transfer for producing 6 transgenic pigs [109] Some other groups have confirmed ICSI-SMGT as a valuable methodology to obtain transgenic embryos and piglets [57, 110-112] and they have investigated factors that affect the efficiency of blastocyst development and protein (EGFP) expression in porcine embryos following ICSI-SMGT. In this way, we have evaluated the effect of the sperm treatments (freezing with and without cryoprotectant agents or Triton x-100) on the efficiency for producing EGFP-expressing pig embryos by ICSI-SMGT. We demonstrated that the integrity of the sperm plasma membrane plays a critical role in DNA interaction, and altered plasma membranes facilitate interactions between an injected exogenous DNA and the sperm chromatin. However, severe sperm treatments may damage the sperm nucleus, induce DNA fragmentation and/or lead to chromosomal breakage with a detrimental effect on further embryonic development [65].

On the other hand we have explored the use of Recombinase-A protein (RecA) coated exogenous DNA that has been used previously in pronuclear injection systems increasing integration into goat and pig genomes [113]. We confirmed that the presence of exogenous DNA and RecA-DNA complexes did not affect sperm functionality in terms of motility, viability, membrane lipid disorder or ROS generation. EGFP-expressing embryos were obtained with a high efficiency using the SMGT-ICSI technique in combination with recombinase although the use of the IVF system did not result in any fluorescent embryos, and finally transgenic piglets were produced by the ICSI-SMGT methodology [108, 114]. To our knowledge, this is the first time that transgenic pigs have been produced by this "active transgenic approach". These findings open up new possibilities in the future for production of transgenic pigs.

OTHER FARM SPECIES

The use of SMGT in others farm species is limited [2, 26]. Although the first studies of DNA binding to sheep and goat sperm were developed in the early 90s [46], these species have been not studied deeply. Some groups have explored the use of TMGT in goats [26, 115, 116] and recently, some studies have been carried out in goats analyzing the DNA binding capacity of the sperm from different bucks [117, 118], subsequently reporting the production of transgenic goats by artificial insemination with fresh and frozen spermatozoa incubated with exogenous DNA [119].

In horses the uptake of exogenous DNA by equine spermatozoa have been demonstrated [120] and shown to be increased by the use of liposomes [121]. However, there was no evidence of expression of EGFP in equine blastocysts after SMGT by artificial insemination [120, 121], but it is possible to produce equine embryos expressing EGFP by ICSI [57].

Table 6: Porcine Sperm-DNA Interaction and Sperm-Mediated Gene Transfer Experiments

Year	Reference	Experimental	Characteristics
1989	Gandolfi *et al.* [83]	AI	Transgenic piglets
1990	Castro *et al.* [46]	DNA incubation	Radio labeling
1991	Horan *et al.* [97]	DNA incubation	Percoll gradient
1992	Horan *et al.* [98]	IVF	No transgenic embryos
1992	Horan *et al.* [100]	DNA incubation	Electroporation
1992	Camanioni *et al.* [48]	DNA incubation	Seminal plasma inhibitory
1996	Sperandio *et al.* [53]	AI	Transgenic embryos and piglets (CAT)
1996	Gandolfi *et al.* [84]	AI	No transgenic piglets
1997	Lavitrano *et al.*[85]	AI	Transgenic embryos and piglets (hDAF)
1999	Lavitrano *et al.* [86]	AI	Transgenic piglets (hDAF)
2000	Sciamanna *et al.* [88]	AI	Retrotransposón Transgenic piglets (hGH)
2001	Lai *et al.* [101]	IVF ICSI	Liposomes Transgenic embryos by ICSI
2001	Kurome *et al.* [106]	ICSI	Transgenic embryos (GFP)
2002	Lavitrano *et al.*[89]	AI	Transgenic piglets (hDAF)
2002	Chang *et al.* [102]	AI	Antibody linker Transgenic piglets (SEAP-2)
2003	Lavitrano *et al.*[70]	AI	Transgenic piglets (hDAF)
2005	Fantinati *et al.* [90]	AI	Laparoscopic AI Transgenic embryos (EGFP)
2005	Webster *et al.* [91]	AI	Laparoscopic AI Muti-transgenic embryos
2005	Naruse *et al.* [107]	ICSI	Transgenic piglet (hAlb)

Table 6: cont....

2006	Kurome *et al.* [109]	ICSI+NT	Transgenic piglets (hAlb)
2006	Yong *et al.* [110]	ICSI	Transgenic piglets (EGFP)
2006	Manzini *et al.* [93]	AI	Episomal vector Transgenic piglets (EGFP)
2007	Kurome *et al.* [111]	ICSI	Transgenic embryos (EGFP)
2008	Pereyra-Bonnet *et al.* [57]	ICSI	Transgenic embryos (EGFP)
2009	Kim *et al.* [103]	IVF	Magnetic Nanoparticle, Liposome Transgenic embryos (EGFP)
2009	Wu *et al.* [112]	ICSI	Transgenic embryos (EGFP)
2009	Bacci *et al.* [104]	IVF	Transgenic embryos (pGeneGrip)
2009	Garcia-Vazquez *et al.* [65]	ICSI	Transgenic embryos (EGFP)
2009	Garcia-Vazquez *et al.*[66]	DNA incubation	Epidydimal spermatozoa
2010	Garcia-Vazquez *et al.* [96]	AI	No transgenic embryos

CONCLUSIONS

Twenty year after the first references to the use of SMGT in farm animals we can conclude that a great effort has been done to understand the basis of this methodology and fortunately it could offer potential advantages in the future over other methods for inducing transgenesis in domestic livestock. At the present moment it is a methodology in development and some important factors must be analysed to improve the repeatability of the technique and its application in the field.

ACKNOWLEDGEMENT

Grant Support: MEC-FEDER AGL-2009-12512-C02-01.

REFERENCES

[1]   Wheeler MB, Walters EM, Clark SG. Transgenic animals in biomedicine and agriculture: outlook for the future. Anim Reprod Sci 2003; 79: 265-89.

[2]   Niemann H, Kues WA. Transgenic farm animals: an update. Reprod Fertil Dev 2007; 19: 762-70.

[3]   Band MR, Larson JH, Rebeiz M, *et al.* An ordered comparative map of the cattle and human genomes. Genome Res 2000; 10: 1359-68.

[4]   Kappes SM. Utilisation of gene mapping information in livestock animals. Theriogenology 1999; 51: 135-47.

[5]   Niemann H, Kues WA. Transgenic livestock: premises and promises. Anim Reprod Sci 2000 2; 61: 277.

[6]   Salamone D, Baranao L, Santos C, *et al.* High level expression of bioactive recombinant human growth hormone in the milk of a cloned transgenic cow. J Biotechnol 2006; 124: 469-72.

[7]   Yang P, Wang J, Gong G, *et al.* Cattle mammary bioreactor generated by a novel procedure of transgenic cloning for large-scale production of functional human lactoferrin. PLoS One 2008; 3: e3453.

[8]   Echelard Y, Williams JL, Destrempes MM, *et al.* Production of recombinant albumin by a herd of cloned transgenic cattle. Transgenic Res 2009; 18: 361-76.

[9]   Melo EO, Canavessi AMO, Franco MM, Rumpf R. Animal transgenesis: state of the art and applications. J Appl Genet 2007; 48: 47-61.

[10]   Keefer CL. Production of bioproducts through the use of transgenic animal models. Anim Reprod Sci 2004; 83: 5-12.

[11]   Dyck MK, Lacroix D, Pothier F, Sirard MA. Making recombinant proteins in animals-different systems, different applications. Trends Biotechnol 2003; 21: 394-99.

[12]   Dyck MK, Gagne D, Ouellet M, *et al.* Seminal vesicle production and secretion of growth hormone into seminal fluid. Nat Biotechnol 1999; 17: 1087-90.

[13]   Hammer RE, Pursel VG, Rexroad CE, *et al.* Production of Transgenic Rabbits, Sheep and Pigs by Microinjection. Nature 1985; 315: 680-83.

[14]   Brem G, Brenig B, Goodman HM, *et al*. Production of transgenic mice, rabbits and pigs by microinjection into pronuclei. Reprod Domest Anim 1985; 20: 251-52.

[15]   Wall RJ. New gene transfer methods. Theriogenology 2002 1; 57: 189.

[16]   Chan AWS. Transgenic animals: current and alternative strategies. Cloning 1999; 1: 25-46.

[17]   Whitelaw CBA, Lillico SG, King T. Production of transgenic farm animals by viral vector-mediated gene transfer. Reprod Domest Anim 2008; 43: 355-58.

[18]   Cabot RA, Kuhholzer B, Chan AWS, *et al*. Transgenic pigs produced using *in vitro* matured oocytes infected with a retroviral vector. Anim Biotechnol 2001; 12: 205-14.

[19]   Sims M, First NL. Production of calves by transfer of nuclei from cultured inner cell mass cells. Proc Natl Acad Sci USA 1994 21; 91: 6143-47.

[20]   Brackett BG, Baranska W, Sawicki W, Koprowski H. Uptake of heterologous genome by mammalian spermatozoa and its transfer to ova through fertilisation. Proc Natl Acad Sci USA 1971; 68: 353.

[21]   Lavitrano M, Camaioni A, Fazio VM, *et al*. Sperm cells as vectors for introducing foreign DNA into eggs: genetic transformation of mice. Cell 1989 2; 57: 717-23.

[22]   Smith K, Spadafora C. Sperm-mediated gene transfer: applications and implications. BioEssays 2005; 27: 551-62.

[23]   Gandolfi F. Spermatozoa, DNA binding and transgenic animals. Transgenic Res 1998; 7: 147-55.

[24]   Spadafora C. Sperm-mediated gene transfer: mechanisms and implications. Soc Reprod Fertil 2007; 65: 459-67.

[25]   Lavitrano M, Busnelli M, Cerrito MG, *et al*. Sperm-mediated gene transfer. Reprod Fertil Devel 2006; 18: 19.

[26]   Niu YD, Liang SL. Progress in gene transfer by germ cells in mammals. J Genet Genomics 2008; 35: 701-14.

[27]   Coward K, Kubota H, Parrington J. *In vivo* gene transfer into testis and sperm: developments and future application. Arch Androl 2007; 53: 187-97.

[28]   Parrington J, Coward K, Hibbitt O, *et al*. *In vivo* gene transfer into the testis by electroporation and viral infection-a novel way to study testis and sperm function. Soc Reprod Fertil 2007; 65: 469-74.

[29]   Kubota H, Hayashi Y, Kubota Y, Coward K, Parrington J. Comparison of two methods of *in vivo* gene transfer by electroporation. Fertil Steril 2005; 83: 1310-18.

[30]   Roschlau K, Rommel P, Andreewa L, *et al*. Gene transfer experiments in cattle. J Reprod Fertil 1989; 38: 153-60.

[31]   Krimpenfort P, Rademakers A, Eyestone W, *et al*. Generation of transgenic dairy cattle using '*in vitro*' embryo production. Biotechnology (NY) 1991; 9: 844-47.

[32]   Seidel GE, Jr. Resource requirements for transgenic livestock research. J Anim Sci 1993; 71: 26-33.

[33]   Hodges CA, Stice SL. Generation of bovine transgenics using somatic cell nuclear transfer. Reprod Biol Endocrinol 2003 7; 1: 81.

[34]   Notarianni E, Galli C, Laurie S, Moor RM, Evans MJ. Derivation of pluripotent, embryonic cell lines from the pig and sheep. J Reprod Fertil 1991; 43: 255-60.

[35]   Piedrahita JA, Anderson GB, Bondurant RH. On the isolation of embryonic stem cells : Comparative behavior of murine, porcine and ovine embryos. Theriogenology 1990; 34: 879-01.

[36]   Piedrahita JA, Moore K, Lee C, *et al*. Advances in the generation of transgenic pigs via embryo-derived and primordial germ cell-derived cells. J Reprod Fertil 1997; 52: 245-54.

[37]   Ezashi T, Telugu BP, Alexenko AP, *et al*. Derivation of induced pluripotent stem cells from pig somatic cells. Proc Natl Acad Sci USA 2009; 106: 10993-98.

[38]   Cibelli JB, Stice SL, Golueke PJ, *et al*. Cloned transgenic calves produced from nonquiescent fetal fibroblasts. Science 1998; 280: 1256-68.

[39]   Cibelli JB, Stice SL, Golueke PJ, *et al*. Transgenic bovine chimeric offspring produced from somatic cell-derived stem-like cells. Nat Biotechnol 1998; 16: 642-6.

[40]   Brophy B, Smolenski G, Wheeler T, *et al*. Cloned transgenic cattle produce milk with higher levels of beta-casein and kappa-casein. Nat Biotechnol 2003; 21: 157-62.

[41]   Solter D. Cloning claims challenged. Nature 1999; 399: 13.

[42]   Campbell KH, Fisher P, Chen WC, *et al*. Somatic cell nuclear transfer: Past, present and future perspectives. Theriogenology 2007; 68: S214-31.

[43]   Galli C, Lagutina I, Lazzari G. Introduction to cloning by nuclear transplantation. Cloning Stem Cells 2003; 5: 223-32.

[44]   Arat S, Gibbons J, Rzucidlo SJ, *et al*. *In vitro* development of bovine nuclear transfer embryos from transgenic clonal lines of adult and fetal fibroblast cells of the same genotype. Biol Reprod 2002; 66: 1768-74.

[45]   Arat S, Rzucidlo SJ, Gibbons J, Miyoshi K, Stice SL. Production of transgenic bovine embryos by transfer of transfected granulosa cells into enucleated oocytes. Mol Reprod Dev 2001; 60: 20-26.

[46] Castro FO, Hernandez M, Uliver C, *et al*. Introduction of foreign DNA into the spermatozoa of farm animals. Theriogenology 1990; 34: 1099-10.

[47] Atkinson PW, Hines ER, Beaton S, *et al*. Association of exogenous DNA with cattle and insect spermatozoa *in vitro*. Mol Reprod Dev 1991; 29: 1-5.

[48] Camaioni A, Russo MA, Odorisio T, *et al*. Uptake of exogenous DNA by mammalian spermatozoa: specific localisation of DNA on sperm heads. J Reprod Fertil 1992; 96: 203-12.

[49] Schellander K, Peli J, Schmoll F, Brem G. Artificial insemination in cattle with DNA-treated sperm. Anim Biotechnol 1995; 6: 41-50.

[50] Alderson J, Wilson B, Laible G, Pfeffer P, L'Huillier P. Protamine sulfate protects exogenous DNA against nuclease degradation but is unable to improve the efficiency of bovine sperm mediated transgenesis. Anim Reprod Sci 2006; 91: 23-30.

[51] Anzar M, Buhr MM. Spontaneous uptake of exogenous DNA by bull spermatozoa. Theriogenology 2006 1; 65: 683-90.

[52] Gagne MB, Pothier F, Sirard MA. Electroporation of bovine spermatozoa to carry foreign DNA in oocytes. Mol Reprod Dev 1991; 29: 6-15.

[53] Sperandio S, Lulli V, Bacci ML, *et al*. Sperm mediated gene transfer in bovine and swine species. Anim Biotechnol 1996; 7: 59-77.

[54] Rieth A, Pothier F, Sirard MA. Electroporation of bovine spermatozoa to carry DNA containing highly repetitive sequences into oocytes and detection of homologous recombination events. Mol Reprod Dev 2000; 57: 338-45.

[55] Shemesh M, Gurevich M, Harel-Markowitz E, *et al*. Gene integration into bovine sperm genome and its expression in transgenic offspring. Mol Reprod Dev 2000; 56: 306-08.

[56] Hoelker M, Mekchay S, Schneider H, *et al*. Quantification of DNA binding, uptake, transmission and expression in bovine sperm mediated gene transfer by RT-PCR: effect of transfection reagent and DNA architecture. Theriogenology 2007; 67: 1097-07.

[57] Pereyra-Bonnet F, Fernandez-Martin R, Olivera R, *et al*. A unique method to produce transgenic embryos in ovine, porcine, feline, bovine and equine especies. Reprod Fert Develop 2008; 20: 741-49.

[58] Perez A, Solano R, Castro FO, *et al*. Sperm cell mediated gene transfer in cattle. Biotecnologia Aplicada 1991; 8: 90-94.

[59] Sirisathien S, Keskintepe L, Bracket BG. Bull Sperm Uptake of Exogenous DNA and Efforts to Obtain Transgenic Embryos. Transgenics 2003; 4: 65-76.

[60] Wang HJ, Lin AX, Chen YF. Association of rabbit sperm cells with exogenous DNA. Anim Biotechnol 2003; 14: 155-65.

[61] Flesch FM, Gadella BM. Dynamics of the mammalian sperm plasma membrane in the process of fertilisation. Biochim Biophys Acta Rev Biomembr 2000; 1469: 197-35.

[62] Canovas S, Gutierrez-Adan A, Gadea J. Heparin increases the dna-binding capacity of frozen/thawed bovine spermatozoa. Reprod Domest Anim 2008; 43: 55.

[63] Zani M, Lavitrano M, French D, *et al*. The mechanism of binding of exogenous DNA to sperm cells: factors controlling the DNA uptake. Exp Cell Res 1995; 217: 57-64.

[64] Canovas S, Gutierrez-Adan A, Gadea J. Bovine sperm mediated gene transfer: use of flow cytometer to evaluate binding exogenous DNA to sperm and its further viability. Reprod Domest Anim 2008; 43: 203-04.

[65] Garcia-Vazquez FA, Garcia-Rosello E, Gutierrez-Adan A, Gadea J. Effect of sperm treatment on efficiency of EGFP-expressing porcine embryos produced by ICSI-SMGT. Theriogenology 2009; 72: 506-18.

[66] Garcia-Vazquez FA, Gutierrez-Adan A, Gadea J. Evaluation of binding sperm-exogenous DNA in ejaculate and epididimary porcine spermatozoa. Arch Med Vet 2009; 41: 131-38.

[67] Garcia-Vazquez F, Gumbao D, Gutierrez-Adan A, Gadea J. Porcine sperm mediated gene transfer: use of flow cytometry to evaluate binding of exogenous DNA to spermatozoa. Reprod Domest Anim 2005; 40: 339-40.

[68] Garcia-Vazquez F, Gumbao D, Gutierrez-Adan A, Gadea J. Use of flow cytometry to evaluate the capacity of boar sperm to bind to exogenous DNA of different sizes. Reprod Fertil Dev 2007; 19: 316.

[69] Kang JH, Hakimov H, Ruiz A, *et al*. The negative effects of exogenous DNA binding on porcine spermatozoa are caused by removal of seminal fluid. Theriogenology 2008; 70: 1288-96.

[70] Lavitrano M, Forni M, Bacci ML, *et al*. Sperm mediated gene transfer in pig: Selection of donor boars and optimisation of DNA uptake. Mol Reprod Dev 2003; 64: 284-91.

[71] Celebi C, Guillaudeux T, Auvray P, Vallet-Erdtmann V, Jegou B. The making of "transgenic spermatozoa". Biol Reprod 2003; 68: 1477-83.

[72]   Feitosa WB, Milazzotto MP, Simoes R, *et al*. Bovine sperm cells viability during incubation with or without exogenous DNA. Zygote 2009 16: 1-20.

[73]   Canovas S, Gutierrez-Adan A, Gadea J. Presence of exogenous DNA reduces the viability and motility of frozen thawed bovine spermatozoa. Reprod Domest Anim 2008; 43: 203.

[74]   Maione B, Pittoggi C, Achene L, Lorenzini R, Spadafora C. Activation of endogenous nucleases in mature sperm cells upon interaction with exogenous DNA. DNA and cell biology 1997; 16: 1087-97.

[75]   Fisher J, Bartoov B. DNase-Ii in bull and ram sperm tail and mitochondria. Arch Androl 1980; 4: 157-70.

[76]   You J, Kamihira M, Iijima S. Enhancement of transfection efficiency by protamine in DDAB lipid vesicle-mediated gene transfer. J Biochem 1999; 125: 1160-67.

[77]   Perry AC, Wakayama T, Kishikawa H, *et al*. Mammalian transgenesis by intracytoplasmic sperm injection. Science 1999 14; 284: 1180-83.

[78]   Zi XD, Chen SW, Liang GN, *et al*. The effect of retroviral vector on uptake of human lactoferrin DNA by Yak (Bos Grunniens) spermatozoa and their fertilisability *in vitro*. Anim Biotechnol 2009; 20: 247-51.

[79]   Gadea J, Garcia Vazquez F. Applications of transgenic pigs in biomedicine and animal production. Informacion Tecnica Economica Agraria 2010; 106: 30-45.

[80]   Niemann H, Kues WA. Application of transgenesis in livestock for agriculture and biomedicine. Anim Reprod Sci 2003 15; 79: 291-17.

[81]   Sachs DH, Galli C. Genetic manipulation in pigs. Current Opinion In Organ Transplantation 2009; 14: 148-53.

[82]   Gadea J, Garcia Vazquez F. Methodologies for generating transgenic pigs. Informacion Tecnica Economica Agraria 2010; 106: 15-29.

[83]   Gandolfi F, Lavitrano M, Camaioni A, *et al*. The use of sperm-mediated gene transfer for the generation of transgenic pigs. J Reprod Fertil 1989; 4: 10.

[84]   Gandolfi F, Terqui M, Modina S, *et al*. Failure to produce transgenic offspring by intra-tubal insemination of gilts with DNA-treated sperm. Reprod Fertil Devel 1996; 8: 1055-60.

[85]   Lavitrano M, Forni M, Varzi V, *et al*. Sperm-mediated gene transfer: production of pigs transgenic for a human regulator of complement activation. Transplant Proc 1997; 29: 3508-09.

[86]   Lavitrano M, Stoppacciaro A, Bacci ML, *et al*. Human decay accelerating factor transgenic pigs for xenotransplantation obtained by sperm-mediated gene transfer. Transplant Proc 1999; 31: 972-74.

[87]   Lazzereschi D, Forni M, Cappello F, *et al*. Efficiency of transgenesis using sperm-mediated gene transfer: generation of hDAF transgenic pigs. Transplant Proc 2000; 32: 892-94.

[88]   Sciamanna I, Piccoli S, Barberi L, *et al*. DNA dose and sequence dependence in sperm-mediated gene transfer. Mol Reprod Dev 2000; 56: 301-05.

[89]   Lavitrano M, Bacci ML, Forni M, *et al*. Efficient production by sperm-mediated gene transfer of human decay accelerating factor (hDAF) transgenic pigs for xenotransplantation. Proc Natl Acad Sci USA 2002 29; 99: 14230-35.

[90]   Fantinati P, Zannoni A, Bernardini C, *et al*. Laparoscopic insemination technique with low numbers of spermatozoa in superovulated prepuberal gilts for biotechnological application. Theriogenology 2005; 63: 806-17.

[91]   Webster NL, Forni M, Bacci ML, *et al*. Multi-transgenic pigs expressing three fluorescent proteins produced with high efficiency by sperm mediated gene transfer. Mol Reprod Dev 2005; 72: 68-76.

[92]   Giovannoni R, Vargiolu A, Bacci ML, *et al*. Production of multi-gene transgenic pigs by sperm-mediated gene transfer for xenotransplantation. Xenotransplantation 2009; 16: 376.

[93]   Manzini S, Vargiolu A, Stehle IM, *et al*. Genetically modified pigs produced with a nonviral episomal vector. Proc Natl Acad Sci USA 2006 21; 103: 17672-77.

[94]   Wu Z, Li Z, Yang J. Transient transgene transmission to piglets by intrauterine insemination of spermatozoa incubated with DNA fragments. Mol Reprod Dev 2008; 75: 26-32.

[95]   Garcia-Vazquez F, Grullon L, Ruiz S, Gutierrez-Adan A, Gadea J. Spermatozoids as exogenous DNA vectors : production of transgenic porcine embryos through different spermatic treatments. Informacion Tecnica Economica Agraria 2007: 54-56.

[96]   Garcia-Vazquez FA, Ruiz S, Grullon LA, *et al*. Sperm mediated gene transfer in pigs: effect of exogenous DNA presence in seminal quality and evaluation of *in vivo* transgenic embryo production. Rev Fac Cienc Vet La Plata Univ Nac La Plata 2010; 20: 81-88.

[97]   Horan R, Powell R, McQuaid S, Gannon F, Houghton JA. Association of foreign DNA with porcine spermatozoa. Arch Androl 1991; 26: 83-92.

[98]   Horan R, Powell R, Gannon F, Houghton JA. The fate of foreign DNA associated with pig sperm following the *in vitro* fertilisation of zona-free hamster ova and zona-intact pig ova. Arch Androl 1992; 29: 199-06.

[99] Carballada R, Esponda P. Regulation of foreign DNA uptake by mouse spermatozoa. Exp Cell Res 2001 15; 262: 104-13.

[100] Horan R, Powell R, Bird JM, Gannon F, Houghton JA. Effects of electropermeabilisation on the association of foreign DNA with pig sperm. Arch Androl 1992; 28: 105.

[101] Lai L, Sun Q, Wu G, *et al*. Development of porcine embryos and offspring after intracytoplasmic sperm injection with liposome transfected or non-transfected sperm into *in vitro* matured oocytes. Zygote 2001; 9: 339-46.

[102] Chang K, Qian J, Jiang M, *et al*. Effective generation of transgenic pigs and mice by linker based sperm-mediated gene transfer. BMC biotechnol 2002 19; 2: 5.

[103] Kim TS, Lee SH, Gang GT, *et al*. Exogenous DNA uptake of boar spermatozoa by a magnetic nanoparticle vector system. Reprod Domest Anim 2010; 45: 201-06.

[104] Bacci ML, Zannoni A, De Cecco M, *et al*. Sperm-mediated gene transfer-treated spermatozoa maintain good quality parameters and *in vitro* fertilisation ability in swine. Theriogenology 2009 1163-70.

[105] Chan AW, Luetjens CM, Dominko T, *et al*. TransgenICSI reviewed: foreign DNA transmission by intracytoplasmic sperm injection in rhesus monkey. Mol Reprod Dev 2000; 56: 325-29.

[106] Kurome M, Wako N, Ochiai T, *et al*. Expression of GFP gene introduced into porcine *in vitro* matured oocytes by intracytoplasmic sperm injection. Transgenic animal research conference III 2001.

[107] Naruse K, Ishikawa H, Kawano HO, *et al*. Production of a transgenic pig expressing human albumin and enhanced green fluorescent protein. J Reprod Devel 2005; 51: 539-46.

[108] Garcia-Vazquez F, Ruiz S, Grullon L, *et al*. Birth of transgenic piglets using SMGT-ICSI technique in combination with RecA recombinase. Reprod Domest Anim 2008; 43: 47-48.

[109] Kurome M, Ueda H, Tomii R, Naruse K, Nagashima H. Production of transgenic-clone pigs by the combination of ICSI-mediated gene transfer with somatic cell nuclear transfer. Transgenic Res 2006; 15: 229-40.

[110] Yong HY, Hao Y, Lai L, *et al*. Production of a transgenic piglet by a sperm injection technique in which no chemical or physical treatments were used for oocytes or sperm. Mol Reprod Dev 2006; 73: 595-99.

[111] Kurome M, Saito H, Tomii R, *et al*. Effects of sperm pretreatment on efficiency of ICSI-mediated gene transfer in pigs. J Reprod Dev 2007; 53: 1217-23.

[112] Wu Y, Liu CJ, Wan PC, Hao ZD, Zeng SM. Factors affecting the efficiency of producing porcine embryos expressing enhanced green fluorescence protein by ICSI-mediated gene transfer method. Anim Reprod Sci 2009; 113: 156-66.

[113] Maga EA, Sargent RG, Zeng H, *et al*. Increased efficiency of transgenic livestock production. Transgenic Res 2003; 12: 485-96.

[114] Garcia-Vazquez FA, Gutierrez-Adan A, Gadea J. Evaluation of the sperm-mediated gene transfer (SMGT) technique by *in vitro* fertilisation in pigs using RecA protein. Reprod Fertil Dev 2008; 20: 230-31.

[115] Gao H-y, Cao Y, Li S-H, Ren Y, Li Q-W. Research on spermatozoa-mediated gene transfer to product transgenic animals. Yichuan 2003; 25: 283-90.

[116] Honaramooz A, Megee S, Zeng WX, *et al*. Adeno-associated virus (AAV)-mediated transduction of male germ line stem cells results in transgene transmission after germ cell transplantation. FASEB J 2008; 22: 374-82.

[117] Zhao Y-J. Effect of sperm parameters on exogenous DNA binding to goat sperm. Chin J Zool 2009; 44: 141-45.

[118] Ye H-H, Dong G, Yuan J-F, *et al*. Exogenous DNA influencing fertilisation of goat sperm cells and expression in early embryos. Yichuan 2008; 30: 1421-26.

[119] Zhao YJ, Wei H, Wang Y, *et al*. Production of transgenic goats by sperm-mediated exogenous DNA transfer method. Asian-Australas J Anim Sci 2010; 23: 33-40.

[120] Ball BA, Sabeur K, Allen WR. Uptake of exogenous DNA by equine spermatozoa and applications in sperm-mediated gene transfer. Anim Reprod Sci 2006; 94: 115-16.

[121] Ball BA, Sabeur K, Allen WR. Liposome-mediated uptake of exogenous DNA by equine spermatozoa and applications in sperm-mediated gene transfer. Equine Vet J 2008; 40: 76-82.

[122] Francolini M, Lavitrano M, Lamia CL, *et al*. Evidence for nuclear internalisation of exogenous DNA into mammalian sperm cells. Mol Reprod Dev 1993; 34: 133-36.

[123] Rottmann O, Antes R, Hofer P, *et al*. Liposome-mediated gene transfer via sperm cells. High transfer efficiency and persistence of transgenes by use of liposomes and sperm cells and a murine amplification element. J Anim Breed Genet 1996; 113: 401-11.

[124] Chan AW, Luetjens CM, Dominko T, *et al*. Foreign DNA transmission by ICSI: injection of spermatozoa bound with exogenous DNA results in embryonic GFP expression and live rhesus monkey births. Mol Hum Reprod 2000; 6: 26.

[125] Petters RM, Shuman RM, Johnson BH, Mettus RV. Gene-transfer in swine embryos by injection of cells infected with retrovirus vectors. J Exp Zool 1987; 242: 85-88.

[126] Kubisch HM, Larson MA, Funahashi H, Day BN, Roberts RM. Pronuclear Visibility, development and transgene expression in IVM/IVF-derived porcine embryos. Theriogenology 1995; 44: 391-01.

[127] Park KW, Cheong HT, Lai LX, *et al*. Production of nuclear transfer-derived swine that express the enhanced green fluorescent protein. Anim Biotechnol 2001; 12: 173-74.

[128] Lai LX, Kolber-Simonds D, Park KW, *et al*. Production of alpha-1,3-galactosyltransferase knockout pigs by nuclear transfer coning. Science 2002; 295: 1089-92.

[129] Whitelaw CBA, Radcliffe PA, Ritchie WA, *et al*. Efficient generation of transgenic pigs using equine infectious anaemia virus (EIAV) derived vector. FEBS Lett 2004; 571: 233-36.

[130] Petters RM, Alexander CA, Wells KD, *et al*. Genetically engineered large animal model for studying cone photoreceptor survival and degeneration in retinitis pigmentosa. Nat Biotechnol 1997; 15: 965-70.

[131] Hao YH, Yong HY, Murphy CN, *et al*. Production of endothelial nitric oxide synthase (eNOS) over-expressing piglets. Transgenic Res 2006; 15: 739-50.

[132] Rogers CS, Stoltz DA, Meyerholz DK, *et al*. Disruption of the CFTR gene produces a model of cystic fibrosis in newborn pigs. Science 2008; 321: 1837-41.

[133] Umeyama K, Watanabe M, Saito H, *et al*. Dominant-negative mutant hepatocyte nuclear factor 1 alpha induces diabetes in transgenic-cloned pigs. Transgenic Res 2009; 18: 697-06.

[134] Kragh PM, Nielsen AL, Li J, *et al*. Hemizygous minipigs produced by random gene insertion and handmade cloning express the Alzheimer's disease-causing dominant mutation APPsw. Transgenic Res 2009; 18: 545-58.

[135] Swanson ME, Martin MJ, Odonnell JK, *et al*. Production of functional human hemoglobin in transgenic swine. Bio-Technology 1992; 10: 557-59.

[136] Velander WH, Johnson JL, Page RL, *et al*. High-level expression of a heterologous protein in the milk of transgenic swine using the cDNA-encoding human protein-C. Proc Natl Acad Sci USA 1992; 89: 12003-07.

[137] Paleyanda RK, Velander WH, Lee TK, *et al*. Transgenic pigs produce functional human factor VIII in milk. Nat Biotechnol 1997; 15: 971-75.

[138] Van Cott KE, Butler SP, Russell CG, *et al*. Transgenic pigs as bioreactors : a comparison of gamma-carboxylation of glutamic acid in recombinant human protein C and factor IX by the mammary gland. Genet Anal Biomol Eng 1999; 15: 155-60.

[139] Park JK, Lee YK, Lee P, *et al*. Recombinant human erythropoietin produced in milk of transgenic pigs. J Biotechnol 2006; 122: 362-71.

[140] Lee HG, Lee HC, Kim SW, *et al*. Production of recombinant human von Willebrand factor in the milk of transgenic pigs. J Reprod Devel 2009; 55: 484-90.

[141] Pursel VG, Pinkert CA, Miller KF, *et al*. Genetic engineering of livestock. Science 1989; 244: 1281-88.

[142] Muller M, Brenig B, Winnacker EL, Brem G. Transgenic pigs carrying cDNA copies encoding the murine Mx1 protein which confers resistance to influenza-virus infection. Gene 1992; 121: 263-70.

[143] Bleck GT, White BR, Miller DJ, Wheeler MB. Production of bovine alpha-lactalbumin in the milk of transgenic pigs. J Anim Sci 1998; 76: 3072-78.

[144] Pursel VG, Wall RJ, Mitchell AD, *et al*. Expression of insulin-like growth factor-I in skeletal muscle of transgenic swine. Transgenic Anim Agric 1999: 131-44.

[145] Golovan SP, Meidinger RG, Ajakaiye A, *et al*. Pigs expressing salivary phytase produce low-phosphorus manure. Nat Biotechnol 2001; 19: 741-45.

[146] Saeki K, Matsumoto K, Kinoshita M, *et al*. Functional expression of a Delta 12 fatty acid desaturase gene from spinach in transgenic pigs. Proc Natl Acad Sci USA 2004; 101: 6361-66.

[147] Lai LX, Kang JX, Li RF, *et al*. Generation of cloned transgenic pigs rich in omega-3 fatty acids. Nat Biotechnol 2006; 24: 435-36.

[148] Effect of Exogenous DNA on Bovine Sperm Functionality Using the Sperm Mediated Gene Transfer (SMGT) Technique." Molecular Reproduction and Development 77(8): 687-698

CHAPTER 9

Manipulation of Sperm for Efficient Production of Transgenic Calves and Chicks

Mordechai Shemesh[1,*], Laurence Shore[1], Yehuda Stram[1], Eliane Harel-Markowitz[1] and Michael Gurevich[2]

[1]*Kimron Veterinary Institute, Israel and* [2]*Tel HaShomer Hospital, Israel*

Abstract: The current method of micromanipulation used for domestic animals results in less than 1% transgenic animals. This makes it extremely difficult to produce transgenic cows and is not feasible for producing transgenic chickens. The purpose of this work was to find a more efficient method for producing transgenic calves and chicks using a combination of two techniques, lipofection and restriction enzyme mediated insertion (REMI). Previously investigators were unable to produce transgenic chickens using lipofection alone. On the other hand, injection of isolated sperm nucleus incubated with restriction enzyme into oocytes has only been shown to be effective in frogs. In this study, we demonstrated for the first time, that lipofection of both DNA and restriction enzyme could be used to successfully integrate DNA into the sperm genome DNA and then used for routine AI to produce transgenic calves and chicks. First it was demonstrated using needle pricking and southern blot analysis of genomic DNA that the restriction enzyme opens up "hot" spots in the sperm genomic DNA. This produces sticky ends by which foreign DNA can be inserted and integrated into the sperm genomic DNA. The "transgenic sperm" thus made were used in IVF and AI to produce embryos expressing a foreign DNA, EGFP (enhanced green fluorescent protein). Using Not I and linearized pEGFP lipofected to sperm for AI resulted with two calves which expressed the exogenous DNA in their lymphocytes as determined by (a) PCR and RT-PCR; (b) specific emission of green fluorescence by the EGFP protein; (c) homology analysis between EGFP DNA and PCR product DNA sequences and (d) Southern blot analysis. Similarly in the chicken, linearized plasmid EGFP sequences with the corresponding restriction enzyme (REMI) were lipofected into the sperm. The transfected sperm were then used for AI in hens and 90% (17/19) of the resultant chicks expressed the exogenous DNA in their lymphocytes as determined by: (a) PCR and RT-PCR; (b) specific emission of green fluorescence by the EGFP; and (c) Southern blot analysis. A complete homology was found between the Jellyfish EGFP DNA and a 313 bp PCR product of DNA from chick blood cells. The procedure was then tested with an additional construct, hFSH. The construct of hFSH consisted of both subunits, α and β and the PCR product used primers for both α and β subunit resulted with a PCR product of 584 bp which was unique to transgenic chickens. The procedure was then used to lipofect a construct of hFSH (Human Follicular Stimulating Hormone) into chicken sperm and used for AI. The resultant offspring were transgenic for at least three generations as determined by: (1) measurement of hFSH protein in chicken blood using enzyme immunoassay and RIA; (2) RT-PCR and PCR; and (3) copy number.

We conclude: (1) that lipofection of both DNA and restriction enzyme into sperm (bovine and chicken) induces the integration of the DNA into the sperm genomic DNA; (2) lipofected sperm can be used in AI to produce a high percentage of transgenic calves and chicks; (3) The integrated gene is expressed in the first, second and third generation; and (4) the method is not limited to specific genes. The technique of lipofection of DNA combined with REMI is therefore an efficient and stable method of producing transgenic domestic animals. Efficient production of transgenic domestic animals could have major impact on gene therapy, improving livestock breeds and the production of valuable pharmaceuticals, *e.g.* hFSH, which could be extracted from eggs and milk.

Keywords: Lipofection, Lipofected sperm, Restriction enzyme, Transgenic chicken, Transgenic cow, hFSH, GFP, pDNA.

INTRODUCTION

Gene manipulation and expression by transgenesis is one the most significant advances made in applied

*Address correspondence to Mordechai Shemesh: Department of Hormone Research, Kimron Veterinary Institute, PO Box 12, 50250 Bet Dagan, Israel; Tel: 972-9-7414969; E-mail: motishemesh@yahoo.com

biology during the last two decades. This approach has enabled scientists to make considerable progress in understanding the genetic and molecular mechanisms involved in normal and pathological developmental process [1-4]. Transgenesis is also a key tool for the development of animals with high added value intended for pharmaceutical or industrial use.

METHODS FOR PRODUCING TRANSGENIC ANIMALS

Pronuclear Microinjection

Among the major techniques that are commonly used in making transgenic animals, pronuclear microinjection was one of the first methodologies shown to be effective, and has come to be used on a wide scale with mice. That this technique could be used to produce transgenic livestock was demonstrated to be feasible over two decades ago [5]. However the method was highly inefficient and costly in the context of agricultural animals. Efficiencies of <1% required the microinjection and transfer of thousands of embryos to produce a few transgenic offspring. A second limitation of the method was the random integration of a highly variable number of copies of the transgenes which resulted in variable expression.

Microinjection was successfully used to generate transgenic cattle [6], goats [7], sheep [8, 9] and pigs [10]. For commercial production of highly valuable human therapeutic agents, there are other methods for transgenesis that have also been shown to be effective in various experiments. Such alternative methods include the use of viral vectors somatic cells clones and sperm mediated gene transfer (SMGT).

Lentiviral Vectors

DNA transfer using viral vectors using lentiviral vectors [11, 12] is currently the most efficient method for producing transgenic animals, including livestock and poultry. However this approach has limitations due to the size of the lentiviral vectors (For review, see [13]).

Somatic Cell Cloning

The second approach involves *in vitro* transfection of embryonic stem cells (ESCs) [14]. The cells are then transferred into a recipient blastocyst before being implanted into the uterus of a pseudopregnant female. It allows the generation of knockout mice by homologous recombination. The targeted gene can then be "turned off" by an appropriate technical procedure such as the Cre-Lox system [15]. However sequential modification must be done in primary cells with finite life spans which is a challenge [16-18]. Moreover, truly totipotent ESCs have not been derived for most animal species. An alternative approach involves the use of nuclear transfer from genetically manipulated somatic cells. This has wider applicability in terms of host range; however, it has low efficiency and is associated with a high level of morbidity and mortality in the first generation animals [17, 18].

Sperm Mediated Gene Transfer (SMGT)

Brackett *et al.* [19] were the first to demonstrate that spermatozoa have the ability to bind exogenous DNA. In 1989, two independent reports made the claim that sperm cells could bind exogenous DNA molecules (transgenes) and transfer these molecules during fertilisation to produce transgenic animals [20, 21]. The approach generated substantial interest because "sperm mediated DNA transfer" was simple and low cost. However, difficulties in reproducibility and low efficiency in integration of transgenes into the animal genome resulted in considerable controversy, which eventually caused a good deal of skepticism of this approach.

Contradictory results have been reported by other investigators working in several mammalian species [22-26]. However successful reports of SMGT have also been published [27]. Perry *et al.* [28] showed that intracytoplasmic injection of "damaged" sperm encoding exogenous DNA can be used to produce transgenic mice.

To improve the efficiency of SMGT, other studies have used enhancement techniques such as electroporation [29] and liposomes [30] to improve the DNA incorporation into the sperm. An elegant method using a linker-based SMGT method was reported in which DNA was mixed with a monoclonal antibody that recognized a surface antigen on the sperm before being incubated with sperm for fertilisation. Using this approach transgenic offspring were produced with high efficiency [31].

In some cases or some species the transgenes may remain extrachromasomal throughout mitosis and meiosis as has been reported recently [32, 33] using zebrafish transfected sperm.

Natural Barriers to SMGT

To protect the genomic species, nature creates barriers against SMGT. Successful SMGT therefore must overcome the barriers. The best documented barriers include an inhibitory factor in seminal fluid of different mammals that prevent the binding of foreign molecules. Zani *et al.* [34] identified such an inhibitory factor (IF-1) in the seminal fluid. An additional barrier is sperm endogenous nuclease which induces the degradation of exogenous DNA [35].

Restriction Enzyme Mediated Integration

A novel method for integrating exogenous DNA into cells is restriction enzyme mediated integration (REMI). REMI utilises a linear DNA that is derived from a plasmid DNA by cutting the plasmid with restriction enzyme to generate single stranded cohesive ends. The linear cohesive ended DNA, together with restriction enzyme, is then introduced into the target cells by lipofection or electroporation [36]. The corresponding restriction enzyme is then thought to cut the genomic DNA at sensitive sites that enable the exogenous DNA to integrate *via* its matching cohesive ends [37]. Kroll and Amaya [38] used REMI to insert exogenous DNA into nuclei isolated from *Xenopus laevis* (African clawed toad) sperm. The isolated nuclei were then manually injected into *Xenopus* oocytes to produce transgenic embryos. REMI was also successfully use in cattle [39, 40] in which it was combined with lipofection to integrate the transgene into the genomic DNA of the sperm before fertilisation. Lipofection alone in the absence of REMI can result in transgenic chickens but the success is low (2/53; [41]). This is particularly important in the chick where the large amount of yolk does not allow for micromanipulation. It has also been adapted to zebrafish [42].

HYPOTHESIS

We hypothesized that avian or bovine DNA-transfected sperm produced by lipofection and REMI would be a highly efficient procedure to produce sperm carrying a transgene. The transformed sperm could then be used in artificial insemination to produce transgenic offspring. This would be a more effective methodology as it would bypass natural barriers to foreign DNA transfer and the simple method of insemination would cause little disruption to the natural processes of fertilisation. Furthermore, facilitating attachment of the foreign DNA with "sticky ends" to the nuclear DNA would decrease random integration.

Integration: Extra- or Intra-Chromosomal?

Exogenous DNA integration in sperm requires bypassing the plasma membrane barrier. The subsequent fate of sperm-bound DNA after delivery into the oocyte remains a contradictory issue; in particular the question of whether foreign molecules of nucleic acid are integrated into the host genome or remain as extra-chromosomal structures is still unresolved. Available data indicate that the fate of exogenous DNA depends on the procedure through which sperm cells interact with the DNA. The generation of non-integrated episomal structures is a highly probable event when foreign DNA molecules are directly incubated with intact spermatozoa that are used in fertilisation assays ([32, 33]; for review see [43]). Integration in the host genome is rare under these conditions although it has been reported by Lavitrano *et al.* [27] and Webster *et al.* [44]. However transmission of non-integrated sequence using the same methods has been reported by the same group [45]. In contrast integrations seem to be favored in the absence of direct interaction between the exogenous DNA molecules and the sperm membrane. A number of procedures have been used to overcome the nature-created barriers to SMGT including: (a) injection of demembranated spermatozoa associated with exogenous DNA molecule into oocytes, or "transgenICSI" [28, 46]; (b) as we have reported, by lipofection to bypass the plasma membrane [39, 47]; and (c) use of antibody directed against specific proteins on the sperm surface. This is a linker mediated method [31]. The available data suggest that the final fate of exogenous DNA molecules depends on whether they interact directly with the membrane or bypass the membrane; if the action is primarily entrance through the membrane, non-integrated extra-chromosomal structures will mostly be generated, whereas integration of the transgene in the genome of sperm is favored in the later [43].

Random Integration

The interaction of sperm cells with exogenous DNA molecules activates endogenous nucleases [35]. These nucleases heavily degrade the foreign DNA. Eventually, they also act locally to cleave chromatin of the sperm nucleus that retains a nucleohistone organization [48-50] The results suggest that discrete sites of nuclease sensitivity exist within the nucleohistone domains of the otherwise tightly packed chromosome of mature sperm cells [51, 52]. These sites are preferential target for the integration of exogenous sequences [53].

Targeted Integration

Liposome exogenous DNA complexes are able to overcome the barrier of cytoplasmic nuclease activity as well as that of the nucleus. It also provides protection from random integration of damaged fragments of DNA. REMI would also reduce random integration as it makes the DNA more "sticky" allowing more rapid integration into sensitive sites.

RESULTS AND DISCUSSIONS

To produce bovine transgenic sperm, REMI in conjunction with lipofection was used and EGFP integration into the sperm genomic DNA was demonstrated by both needle pricking DNA amplification by PCR and by Southern blot analysis. The data presented strongly suggest that EGFP was integrated into a unique site in the sperm cell genome before fertilisation. REMI lipofected sperm cells were then used in IVF to produce morulae expressing EGFP as determined by specific emission of green light and RT-PCR. Bovine sperm lipofected with Not I linearised pEGFP and Not I endonuclease were also used in artificial insemination of cows. Two live calves were produced, both of which demonstrated specific emission of green light. Both calves were positive for EGFP in PCR and RT-PCR, but the presence of EGFP in Southern blot analysis was demonstrated for only one calf, perhaps due to mosaic expression.

We found therefore that the REMI lipofected sperm could be used in both IVF and AI. However, since both claves produced by AI were positive for EGFP, it would appear that AI is superior to IVF for producing transgenic calves. When the same method was applied to chicken, 17/20 of the first generation chicks were positive for EGFP expression and 83% (25/30) of the second generation. It was demonstrated that transgene DNA in the form of EGFP was integrated into the sperm DNA as shown by Southern blot (Fig. **1**).

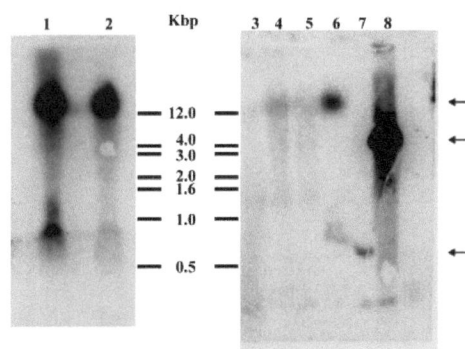

Figure 1: Southern Blot to Determine the Presence of EGFP-DNA in Avian Sperm after Lipofection and REMI.

Lanes 1, 2, 4, 5, 6 – DNA from transfected sperm from five different pools. A fragment of 14 bp indicated that the gene for EGFP was integrated into the genomic DNA of the sperm.

Lane no. 3 – DNA from sperm which underwent transfection in the absence of restriction enzyme (negative control).

Lane no. 7 – pEGFP which was incubated with sperm cells. The DNA of the sperm cells was cut with restriction enzyme *Not I + Hind III* and served as a positive control for free EGFP as a single band of 700 bp was observed.

Lane no. 8 – pEGFP which was linearised with restriction enzyme *Not I* and used a positive control for the plasmid as the fragment obtained was 4.7 bp which is the size of the plasmid EGFP.

The data demonstrate that lipofection of p-DNA in conjunction with restriction enzyme to sperm is a highly efficient method for the production of transfected sperm to produce transgenic offspring by direct AI. The protocol is distinguished by its simplicity in that no special training or equipment is required to execute it and it can be used in animals, like the chicken, where micromanipulation is not feasible [54].

The major question in producing transgenic animals is if the transgene is integrated into the host genome or is episomal. It was expected that the integration would be genomic since we have already shown that in lipofected REMI sperm that the transgene is integrated into the genome as determined by Southern blot. In this study, Southern blot analysis of the transgenic chicken DNA was used to demonstrate integration. Digestion with Xba I resulted in a product of 6 kbp which was recognized by the probe for EGFP. Xba I digestion of the transgenic chicken genome DNA should have yielded a fragment of length size of 4.7 kbp if the DNA was not integrated into genomic DNA, similar to that seen with digestion of the free pEGFP. Therefore the product of 6 kbp observed demonstrated integration of the transgene into chicken genomic DNA. Similarly, the combination of Hind III and Not I, resulting in a free insert of 1.3 kbp, indicated that integration had occurred. Restriction enzyme analysis therefore strongly indicates that only one copy of the EGFP was integrated in one specific site of the genome. The observation of only one site of integration suggests that there are a limited number of susceptible sites on the genome [54].

Sperm mediated gene transfer (SMGT) requires the exogenous DNA to breach a number of barriers. The sperm membrane generally prevents the transfer of DNA as it binds and immobilises the DNA. DNA which does cross the sperm barrier is rarely integrated but episomal transmission is possible as an extra-chromosomal episome. Endogenous nucleases within the cytoplasm can result in a non-specific random integration of damaged fragments of DNA [35]. Finally the DNA needs to find areas susceptible to integration, which are limited. It is possible that the DNA can be transcribed to mRNA, and endogenous RT-transcriptases [43], in turn can convert it to cDNA, but this seems to be a rare occurrence. In the transfected sperm REMI protocol described here, the sperm membrane is bypassed using lipofection and the liposomes provided protection against endogenous nucleases. The restriction enzyme in turn opens up a number of sites to produce "sticky ends" which are then suitable for integration. Lipofection alone in the absence of REMI can result in transgenes in chickens but the success rate is low (2/53) [41]. In contrast, our protocol produced a high percentage of transgenic chickens and the transgene was expressed through three generations. It is anticipated that (all) subsequent generations will retain the transgene. This approach is particularly important for the chicken where the large amount of yolk does not allow for micromanipulation, and transgenic chickens are produced by modification of the blastodermal cells [54]. To demonstrate that the method was valid for DNA other than EGFP, p-hFSH of alpha, beta subunits and leader sequences were lipofected into chicken sperm to produce chickens expressing the hFSH transgene. The PCR fragment product could not be produced naturally in non-trangenic chicks as the alpha and beta subunits and leader sequences are under the control of different genes. Human FSH expression in CHO ovarian cells was demonstrated following lipofection with pFSH (Fig. **2**).

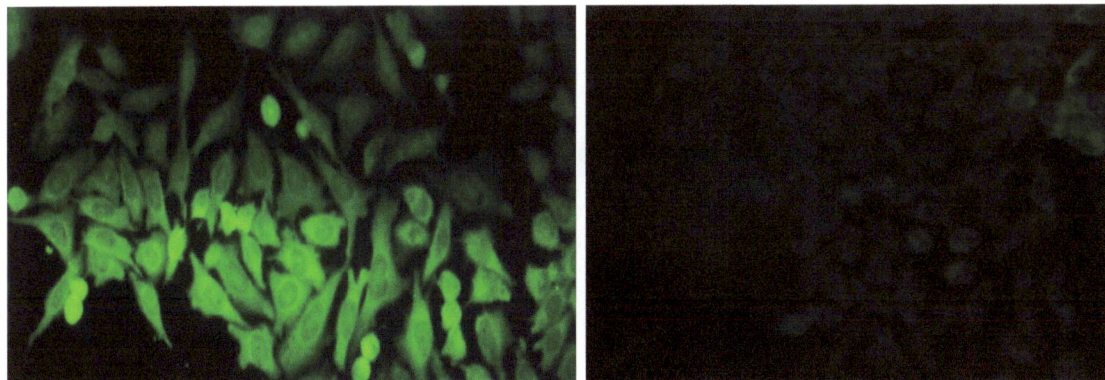

Figure 2: Detection of hFSH Produced in p-hFSH Lipofected CHO Cells.

Cells were fixed in 4% paraformaldehyde, washed and incubated with anti-hFSH primary antibody followed by a second antibody conjugated to FITC. The expression of the hFSH in the cells was determined by immuno-fluorescence using specific antibody against hFSH and a specific green fluorescence appeared solely in the cytoplasm of the lipofected cell. A) p-hFSH lipofected cells. B) Lipofected in the absence of p-hFSH. Original magnification x200.

The sequencing of the PCR product found in the transgenic chickens indicated it was essentially completely homologous with p-hFSH. Furthermore, since a radio-immunoassayable hFSH was found in the blood of the transgenic chickens, it should be possible to produce a mammalian protein in the avian eggs.

Although the data presented are consistent with integration of the EGFP and hFSH sequences into the genome and subsequent expression of the protein by the cell apparatus, several of the observations made were not expected: (1) Why was the expression of EGFP by the chicken cells lower than that seen in transgenic calves? (2) Why was the copy number for hFSH in G2 higher than in G1? And (3) why did the expression of hFSH decrease with age?

Expression of eEGFP by the Chicken Cells

In the transgenic chicken lymphocytes, about 30% of the cells expressed the EGFP. This can be compared to bovine transgenic lymphocytes where the percentage was about 60%. The reason for the failure of the EGFP to be expressed in a higher percentage of the lymphocytes examined is not known. However it could well be that the low percentage observed is due to the sensitivity of the FACS assay. The limit of detection of EGFP (~ 30 μM) in fluorescence microscopy is equivalent to ~4, 000 cytoplasmic molecules per cell [55] and many of the cells may express EGFP below this limit. Alternatively, it could be that the gene may insert into a locus which is activated differently in different lymphocyte types (*e.g.* mononuclear, polynuclear); or the lack of expression in some of the cells could be due to suppressive activity related to specific silent regions in the genome. Another factor may be that the lymphocytes are differentiated stem cells and differentiating cells express the same gene differently as they differentiate, *e.g.* when bovine granulosa cells differentiate into corpora lutea cells, they essentially lose the ability to express the gene for aromatase [56, 57].

Increase in Copy Number in Subsequent Generations

Using quantitative real time PCR it was demonstrated that the transgene copy number was significantly increased in the G2 (produced by interbreeding between transgenic chickens) compared with G1. In the second generation, the gene was expressed in heart, skin and liver by both PCR (Fig. **3**) and quantitative real time PCR (Fig. **4**).

The sample amplification curves begin rising between the 35th and 39th cycle of PCR, while negative control (not shown) remained horizontal. The differences in the cycle no. of the curves represent differences in the gene expression in different tissues.

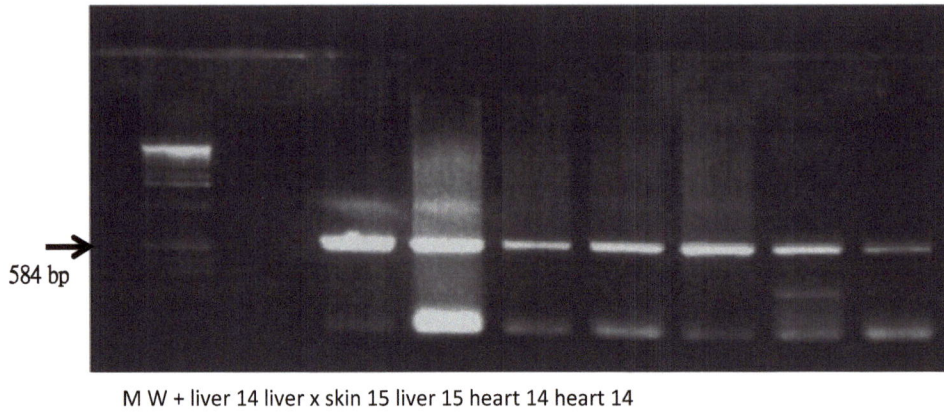

M W + liver 14 liver x skin 15 liver 15 heart 14 heart 14

Figure 3: PCR Product using DNA Extracted from Transgenic Chick Tissues Expressing hFSH. A fragment of 584bp specific for both alpha and beta subunit is demonstrated. M = Molecular markers; W = water; + = positive hFSH control.

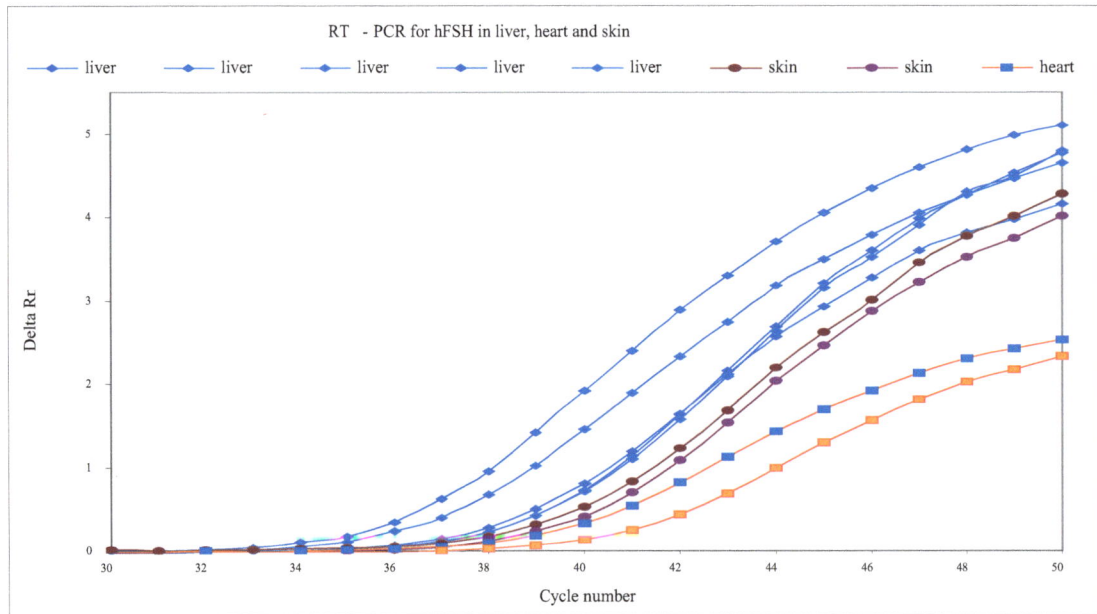

Figure 4: Real Time RT-PCR for hFSH from Second Generation Transgenic Chicken Tissues using 1:10 Serial Dilutions of 0.2 µg/tube of Extracted RNA.

Copy number in the G3 was at least equal to the number in the G2. This can be explained by the integration of plasmid DNA into more than one site in the presence of nonequal sites in the mating chickens of G2 and G3. We therefore hypothesize that in the G1 of transgenic chicks may have had one or more copy of transgene from the transfected sperm. Interbreeding between G1 transgenic chickens would then lead to development of monozygotes or amplification of gene copy numbers in next generations. More importantly, increased copy number of the transgene in G2 and G3 inbred chicken supports our conclusion that true genomic integration occurred as opposed to epigenesis. In case of epigenic transfection, the copy number of the transgene would have tended to be lowered (diluted) in the next generations.

Decrease in the Expression of hFSH with Age

Human FSH protein was demonstrated in the peripheral blood of the G1 and G2. However the level of the hormone in the blood declined to a low level after 1.5 yrs and became undetectable by two years of age.

The loss of transgene expression with aging may be related to time-dependent silencing of CMV promoters, as observed with *in vivo* models. In a gene therapy study, Loser *et al.* [58] demonstrated that hybrid promoter consisting of a minimal CMV promoter and the enhancer II of hepatitis B virus became inactive with age. The reason for the shutdown of the CMV promoter was probably stimuli known to activate the transcription factor NFKB, which binds to four sites in the CMV promoter/enhancer. NFKB is known to be activated by bacterial lipopolysaccharides and it could well be that sporadic infections which occurred during the lifetime of the transgenic chicken could have been responsible for the silencing of the transfected gene with aging.

SMGT as a Method of Overcoming Interspecies Barriers

The methods described provide an important new prospective in the field of animal transgenesis. However, a number of questions remain to be answered. First, large differences in transfection exist between species. Undoubtedly the plasma membrane of the spermatozoa is one of the obstacles preventing the incorporation of exogenous DNA. Therefore, because permeabilisation of this membrane is often required to facilitate foreign DNA transfer, damage incurred to the genomic DNA of the spermatozoon after permeabilisation may in some cases have deleterious effects on the integrity of the generated animals. Furthermore, in most of the studies performed (except with cattle, in which Rieth *et al.* [59] successfully achieve homologous recombination), the transgenes were randomly integrated into the spermatozoa genome, as in the classical technique of additive transgenesis. Therefore REMI is at present the best method available to avoid random integration.

Other important factors that may play a significant role are the insertional mechanisms involved in transgene integration and their effects on DNA structure. During spermatogenesis, the germ cell nucleus undergoes numerous modifications in which the sequential expression of histones, intermediary proteins and protamines ultimately leads to the complex compaction of the chromatin [60]. Condensation of the DNA in spermatozoa shows intra- and inter-species variation which most probably explain, in part, the difference in efficiency observed between techniques from one species to another. Our proposed method also overcome this barrier and reduced the variation.

Transgenesis as a Method for Producing Animals Resistant to Pathogens

The present work was conducted in parallel with efforts to produce transgenic animals resistant to disease. Based on our work where a siRNA was introduced into mammalian cells, demonstrating gene silencing of the corresponding gene - COX-2 [61], methods were developed to inhibit pathogen replication in tissue culture against FMDV and Akabane Viruses [62, 63]. We suggest that the combination of these two technologies, transfected sperm and gene silencing, would be a promising methodology to produce farm animals resistant to different pathogens independent of the immune system.

CONCLUDING REMARKS

We consider that exogenous DNA and restriction enzyme sperm lipofection is a highly efficient transfection procedure. We have demonstrated that lipofected sperm can then be used in AI to produce transgenic offspring. Transgenic chickens expressing both non-mammalian and mammalian genes were produced. The technique is not limited to chicken and cattle as equine embryos [64] carrying EGFP transgenes have also been produced.

REFERENCES

[1]   Chan AWS, Luetjens CM, Dominko T, *et al.* Foreign DNA transmission by ICSI: injection of spermatozoa bound with exogenous DNA results in embryonic EGFP expression and live Rhesus monkey births. Mol Hum Reprod 2000; 6: 26-33.

[2]   Celebi C, Guillaudeux T, Auvray P, *et al.* The making of "transgenic spermatozoa". Biol Reprod 2003; 68: 1477-38.

[3]   Mullins LJ, Mullins JJ. Transgenesis in the rat and larger mammals. J. Clin Invest 1996; 97: 1557-60.

[4]   Wall RJ. New gene transfer methods. Theriogenology 2002; 57: 189-201.

[5]   Hammer RE, Pursel V, Rexroad C, *et al*. Production of transgenic rabbits, sheep and pigs by microinjection. Nature 1985; 315: 680-83.

[6]   Eyestone WH. Production and breeding of transgenic cattle using *in vitro* embryo production technology. Theriogenology 1999; 51: 509-17.

·[7]   Krimpenfort P, Rademakers A, Eyestone W, *et al*. Transgenic production of a variant of human tissue-type plasminogen activator in goat milk: generation of transgenic goats and analysis of expression. Nat Biotechnol 1991; 9: 844-47.

[8]   Ebert KM, Selgrath JP, DiTullio P, *et al*. High-level expression of a heterologous protein in the milk of transgenic swine using the cDNA encoding human protein C. Nat Biotechnol 1991; 9: 835-38.

[9]   Wright G, Carver A, Cottom D, *et al*. A high level expression of active human alpha-1-antitrypsin in the milk of transgenic sheep. Nat Biotechnol 1991; 9: 830-34.

[10]   Velander WH, Johnson JL, Page RL, *et al*. *In vivo* gene delivery and stable transduction of nondividing cells by a lentiviral vector. Proc Natl Acad Sci 1992; 89 : 12003-07.

[11]   Naldini L, Blomer U, Gallay P, *et al*. Efficient transduction of nondividing human cells by feline immunodeficiency virus lentiviral vectors. Science 1996; 272: 263-67.

[12]   Poeschla EM, Wong-Staal F, Looney DJ. Efficient transduction of nondividing human cells by feline immunodeficiency virus lentiviral vectors. Nat Med 1998; 4: 354-57.

[13]   Robl JM, Wang Z, Kasinathan P, Kuroiwa Y. Transgenic animal production and Anim Biotechnol Theriogenol 2007; 67: 127-33.

[14]   Capecchi MR. Altering the genome by homologous recombination. Science 1989; 244: 1288-92.

[15]   Lakso M, Sauer B, Mosinger B Jr, *et al*. Targeted oncogene activation by site specific recombination in transgenic mice. Proc Natl Acad Sci 1992; 89: 6232-36.

[16]   Zwaka TP, Thomson JA. Homologous recombination in human embryonic stem cells. Nat Biotechnol 2003; 21: 319-21.

[17]   Denning C, Priddle H. New frontiers in gene targeting and cloning: success, application and challenges in domestic animals and human embryonic stem cells. Reproduction 2003; 126: 1-11.

[18]   Piedrahita J. Targeted modification of the domestic animal genome. Theriogenology 2000; 53: 105-16.

[19]   Brackett BG, Baranska W, Sawicki W, Koprowski H. Uptake of heterologous genome by mammalian spermatozoa and its transfer to ova through fertilisation. Proc Natl Acad Sci 1971; 68: 353-57.

[20]   Arezzo F. Sea urchin sperm as a vector of foreign genetic information. Cell Biol Int Rep 1989; 13: 391-04.

[21]   Lavitrano M, Camaioni A, Fazio VM, *et al*. Sperm cells as vectors for introducing foreign DNA into eggs: genetic transformation of mice. Cell 1989; 57: 717-23.

[22]   Smith K, Spadafora C. Sperm-mediated gene transfer: applications and implications. BioEssays 2005; 27: 551-62.

[23]   Brinster RL, Sandgren EP, Behringer RR, Palmiter RD. No simple solution for making transgenic mice. Cell 1989; 59: 239-41.

[24]   Hochi S, Ninomiya T, Mizuno A, *et al*. Fate of exogenous DNA carried into mouse eggs by spermatozoa. Anim Biotechnol 1990; 1: 21-31.

[25]   Schellander K, Peli J, Schmoll F, Brem G. Artificial insemination in cattle with DNA-treated sperm. Anim Biotechnol 1995; 6: 41-50.

[26]   Gandolfi F. Sperm-mediated transgenesis. Theriogenology 2000; 53: 127-37 Sperandio S, Lulli V, Bacci ML, *et al*. Sperm-mediated DNA transfer in bovine and swine species. Anim Biotechnol 1996; 7: 59-77.

[27]   Lavitrano M, Bacci ML, Forni M, *et al*. Efficient production by sperm mediated gene transfer of human decay accelerating factor (hDAF) transgenic pigs for xenotransplantation. Proc Natl Acad Sci USA 2002; 99: 14230-35.

[28]   Perry ACF, Wakayama T, Kishikawa H, *et al*. Mammalian transgenesis by intracytoplasmic sperm injection. Science 1999; 284: 1180-83.

[29]   Gagne MB, Pothier F, Sirard MA. Electroporation of bovine spermatozoa to carry foreign DNA in oocytes. Mol Reprod Dev 1991; 29: 6-15.

[30]   Bachiller D, Schellander K, Peli J, Ruther U. Liposome-mediated DNA uptake by sperm cells. Mol Reprod Dev 1991; 30: 194-00.

[31]   Chang K, Qian J, Jiang M, *et al*. Effective generation of transgenic pigs and mice by linker based sperm-mediated gene transfer. BMC Biotechnol 2002; 2: 5.

[32]   Khoo HW, Ang LH, Lim JHB, Wong V. Sperm cells as vectors for introducing foreign DNA into the zebrafish. Aquaculture 1992; 107: 1-19.

[33]   Khoo HW. Sperm-mediated gene transfer studies on zebrafish in Singapore. Mol Reprod Dev 2000; 56: 278-80.

[34]   Zani M, Lavitrano M, French D, *et al.* The mechanism of binding of exogenous DNA to sperm cells-factors controlling the DNA uptake. Exp Cell Res 1995; 217: 57-64.

[35]   Maione B, Pittoggi C, Achene L, *et al.* Activation of endogenous nucleases in mature sperm cells upon interaction with exogenous DNA. DNA Cell Biol 1997; 16: 1087-97.

[36]   Felgner PL, Gadek TR, Holm M, *et al.* Lipofection: a highly efficient, lipid mediated DNA-transfection procedure. Proc Natl Acad Sci USA 1987; 84: 7413-17.

[37]   Kuspa A, Loomis WF. Ordered yeast artificial chromosome clones representing the Dictyostelium discoideum genome. Proc Natl Acad Sci USA 1996 93: 5562-66.

[38]   Kroll K, Amaya E. Transgenic Xenopus embryos from sperm nuclear transplantations reveal FGF signaling requirements during gastrulation. Development 1996; 122: 3173-83.

[39]   Shemesh M, Gurevich M, Harel-Markowitz, E, *et al.* Gene integration into bovine sperm genome and its expression in transgenic offspring. Mol Reprod Devel 2000; 56: 306-08.

[40]   Shemesh M, Gurevich M, Harel-Markowitz E, *et al.* Alternative methods to micromanipulation for producing transgenic cattle. In: Fields MJ, Sand RS, Yelich J, Eds. Factors Affecting Calf Crop: Biotechnology of Reproduction. Boca Raton FL, CRC Press; 2001; pp. 205-18.

[41]   Yang CC, Chang HS, Lin CJ, *et al.* Cock spermatozoa serve as the gene vector for generation of transgenic chicken (Gallus gallus). Asian-australasian J Anim. Sci 2004; 17: 885-91.

[42]   Jesuthasan S, Subburaju S. Gene transfer into zebrafish by sperm nuclear transplantation. Develop Biol 2002; 242: 88-95.

[43]   Spadafora C. Sperm-mediated 'reverse' gene transfer: a role of reverse transcriptase in the generation of new genetic information. Hum Reprod 2008; 23: 735-40.

[44]   Webster NL, Forni M, Bacci ML, *et al.* Multitransgenic pigs expressing three fluorescent proteins produced with high efficiency by sperm mediated gene transfer. Mol Reprod Dev 2005; 72: 68-76.

[45]   Manzini S, Vargiolu A, Stehle IM, *et al.* Genetically modified pigs produced with a nonviral episomal vector. Proc Natl Acad Sci 2006; 103: 17672-77.

[46]   Kurome M., Ueda H, Tomii R, *et al.* Production of transgenic-clone pigs by the combination of ICSI-mediated gene transfer with somatic cell nuclear transfer. Transgenic Res 2006; 15: 229-40.

[47]   Harel-Markowitz E, Gurevich M, Shore LS, *et al.* Use of sperm plasmid DNA lipofection combined with restriction enzyme mediated insertion (REMI) for production of transgenic chickens expressing eEGFP (enhanced Green Fluorescent Protein) or Human Follicle Stimulating Hormone. Biol Reprod 2009; 80: 1046-52

[48]   Gatewood JM, Cook GR, Balhorn R, *et al.* Sequence specific packaging of DNA in human sperm chromatin. Science 1987; 236: 962-64.

[49]   Gatewood JM, Cook GR, Balhorn R, *et al.* Isolation of four core histones from human sperm chromatin represenring a minor subset of somatic histones. J Biol Chem 1990; 265: 20662-66.

[50]   Pittoggi C, Renzi L, Zaccagnini G, *et al.* A fraction of mouse sperm chromatin is organized in nucleosomal hypersensitive domains enriched in retroposon DNA. J Cell Sci 1999; 112: 3537-48.

[51]   Shaman JA, Prisztoka R, Ward WS. Topoisomerase IIB and an extracellular nuclease interact to digest sperm DNA in an apoptotic-like manner. Biol Reprod 2006; 75: 741-48.

[52]   Sotolongo B, Huang TT, Isenberger E, Ward WS. An endogenous nuclease in hamster, mouse, and human spermatozoa cleaves DNA into loop-sized fragments. J Androl 2005; 26: 272-80.

[53]   Pittoggi C, Zaccagnini G, Giordano R, *et al.* Nucleosomal domains of mouse spermatozoa chromatin as potential sites for retroposition and foreign DNA integration. Mol Reprod Dev 2000; 56: 248-51.

[54]   Koo BC, Kwon MS, Choi BR, *et al.* Production of germline transgenic chickens expressing enhanced green fluorescent protein using a MoMLV-based retrovirus vector. FASEB J 2006; 20: 2251-60.

[55]   CLONETECHniques (1997) Technical tips: green fluorescent protein. XII: 22.

[56]   Mason NR, Marsh JM, Savard K. An action of gonadotropin *in vitro.* J Biol Chem 1962; 237: 1801-06. 57.

[57]   Savard K. The biochemistry of the corpus luteum. Biol Reprod 1973; 8: 183- 02.

[58]   Loser P, Jennings G.S, Strauss M, Sandig V. Reactivation of previously silenced cytomegalovirus major immediate-early promoter in the mouse liver: involvement of NFKB. J Virol 1998; 72: 180-90.

[59]   Rieth A, Pothier F, Sirard MA. Electroporation of bovine spermatozoa to carry DNA containing highly repetitive sequences into oocytes and detection of homologous recombination events. Mol Reprod Dev 2000; 57: 338-45.

[60]   Hecht NB. Molecular mechanisms of male germ cell differentiation. Bioessays 1998; 20: 555-61.

[61]   Xiuzhu T, Shore L, Stram Y, Michaeli S, *et al.* Duplexes of 21 nucleotide RNA specific for COX II mediates RNA interference in cultured bovine aortic coronary endothelial cells (BAECs). Prostag Oth Lipid M 2003; 71: 119-29.

[62] Kahana R, Kuznetzova L, Rogel A, *et al.* Inhibition of foot-and-mouth disease virus replication by small interfering RNA. Gen Virol 2004; 85: 3213-17.

[63] Levin A, Kuznetzova L, Kahana R, Rubinstein-Guini M, *et al.* Highly effective inhibition of Akabane virus replication by siRNA genes. Virus Res 2006; 120: 121-27.

[64] Ball BA, Sabeur K, Allen WR. Liposome-mediated uptake of exogenous DNA by equine spermatozoa and applications in sperm-mediated gene transfer. Equine Vet J 2008; 40: 76-82.

CHAPTER 10

The Use of Intracytoplasmic Sperm Injection (ICSI) for Gene Transfer in Mice

Raúl Fernández-González, Pablo Bermejo-Álvarez, Miriam Pérez-Crespo, Alberto Miranda, Ricardo Laguna, Celia Frutos and Alfonso Gutiérrez-Adán[*]

Department of Animal Reproduction, INIA, Ctra de la Coruña KM 5.9, Madrid 28040, Spain

Abstract: Using the mouse model, intracytoplasmic sperm injection (ICSI)-mediated transgenesis has been shown to be a valuable tool for the production of transgenic animals, an essential instrument for basic and applied research in bioscience. This method of transgenesis consists of the microinjection of spermatozoa pre-incubated with foreign DNA. ICSI of DNA-loaded sperm cells has been shown to mediate mouse transgenesis at high efficiency, especially when sperm cell damage (by freeze-thaw cycles or exposure to detergents) is induced. The greatest advantage of ICSI-mediated transgenesis is that it allows introduction of very large DNA transgenes (*e.g.*, yeast artificial chromosomes), with relatively high efficiency into the genome of hosts, as compared to pronuclear microinjection. In addition, ICSI-mediated transgenesis is associated with (a) low frequencies of embryo mosaicism, one of the major limitations of pronuclear microinjection for the production of transgenic livestock, and (b) a high frequency of Mendelian germline transmission of transgenic sequences among founder animals. In this chapter we will review some factors that can increase the efficiency of the ICSI-mediated transgenesis and we will describe a new active form of ICSI-mediated transgenesis employing fresh sperm in conjunction with recombinase or transposase molecules.

Keywords: Intracytoplasmic sperm injection, Mice, Large DNA transgenes, Efficiency, Sperm cell damage, Active transgenesis, Rec A, Mosaicism, DNA integration, Matrix attachment regions, Transmission.

INTRODUCTION

The first indication that exogenous DNA could be introduced into untreated sperm was provided in 1971 [1]. Next, it was reported that live mouse spermatozoa incubated with exogenous DNA could result in transgenic offspring [2]. This pioneer study triggered initial contradictory reports [3] that have fed an intense debate, not yet finished within the field [4-6]. Ten years later, the use of DNA-coated sperm as a vehicle for animal transgenesis was rediscovered; a new reproducible technique involving the previously developed methodology of intracytoplasmic sperm injection (ICSI) [7], was shown to mediate mouse transgenesis at high efficiency [8]. The technique, which has been termed "metaphase II transgenesis" or "transgenICSI", experienced several improvements, such as the use of piezo-actuated micromanipulation, which avoid the high oocyte mortality caused by conventional pipettes, and the optimisation of the regulated damage of sperm heads, necessary for the transgene insertion. Several transgene archetypes were assayed with this method, including some genomic-type transgenes (BAC or small mammalian artificial-chromosomes), ranging from 11.9 to 170 kilobases (kb) [9], thus revitalising the interest in ICSI techniques within the mammalian transgenesis field (see Table **1**, at foot of this chapter). Later, the initially reported MII–ICSI method was extended to larger genomic transgenes, allowing the generation of transgenic mice carrying a 250-kb YAC by ICSI [10], at higher efficiencies (35%) than those reported for pronuclear microinjection methods (< 5%, usually 1% or lower) (Table **1**) [11]. This method is described in detail in this chapter, and its application has been successfully extended to the generation of transgenic mice with larger YAC DNA molecules (520 kb), double in size to those tested before [10], which carried the human amyloid-precursor protein gene (Table **1**). Again, higher efficiencies were observed (10–20%), these being variable and dependant upon the concentration of the YAC DNA samples used [12]. Furthermore, large DNA inserts (*e.g.*, artificial chromosomes) usually include all regulatory elements, and thus frequently

---

*Addres correspondence to Alfonso Gutiérrez-Adán: Department of Animal Reproduction, INIA, Ctra de la Coruña KM 5.9, Madrid 28040, Spain; Tel: 0034 913473768; E-mail: agutierr@inia.es

result in correct gene expression, mimicking the endogenous expression pattern of the homologous locus, regardless of position, unlike the limited promoter-driven cDNA transgenes.

Table 1: Transgenic Efficiency Comparisons of Different Transgenesis Methods

Transgenesis method	Transgene Size Kb	Strain	Oocyte injected (repetitions)	Embryo transferred (recipients)	Births (% transferred) (%oocytes)
Pronuclear	Various	Varius	4739 (NA)	NA	626 (NA)(13.2)
Pronuclear	250	CD1	1733 (12)	959 (44)	93 (10)(5.4)
ICSI-Tr	5	CD1	219 (6)	163 (8)	22 (13) (10)
ICSI-YAC	250	CD1	367 (6)	218 (13)	34 (16)(9.3)
ICSI-YAC	520	B6D2F1	311 (5)	228 (12)	52 (23) (17)
ICSI-Tr	5	B6D2F1	97 (1)	179 (12)	12 (26.6) (12.3)
ICSI-Tr	11-170	B6D2F1	NA	179 (12)	14 (7.8) (NA)
TN:ICSI	3.6	B6D2F1	204 (7)	171 (14)	107 (62.6) (52)
TN:ICSI	3.6	C57BL/6	94 (2)	77 (6)	45 (58.4) (47.9)
ICSI:Lysolecithin	5	B6D2F1	44 (2)	23 (2)	8 (34.7) (18.2)
ICSI-NaOH	5	B6D2F1	163 (5)*	96 NA	26 (27.1) (16)

Transgenesis method	Transgenic pups			Publication reference
	Total	% animals born	% Oocytes injected	
Pronuclear	150	24	3.2	[35]
Pronuclear	1	1	0.06	[10]
ICSI-Tr	10	45	4.6	[10]
ICSI-YAC	12	35	3.3	[10]
ICSI-YAC	11	21	3.5	[12]
ICSI-Tr	2	16.6	2	[8]
ICSI-Tr	6	42.8	NA	[9]
TN:ICSI	23	21.5	11.3	[14]
TN:ICSI	4	8.8	4.3	[14]
ICSI:Lysolecithin	5	62.5	11.4	[31]
ICSI-NaOH	12	46.2	7.4	[29]

*No. surviving oocytes.
ICSI-Tr is done with freeze-thawed sperm and linear transgene. TN:ICSI is achieved with the Tn5 transposase purchased from manufacturer in the protein form and complexed with a EGFP transgene containing transposon. The transposon–transposase complex along with freshly immobilized sperm is microinjected during ICSI. TP:ICSI is performed by microinjecting the pMMK-2 plasmid construct which contains the donor and helper elements with freshly immobilized sperm.

Efficiencies reported refer to DNA-transgene positive founder mice. As in any microinjection technique, not all YAC transgenic mice generated eventually carried the entire YAC DNA molecules, but a fragment of it. However, in two out of 12 transgenic founder mice carrying the 250-kb YAC we could demonstrate the presence of the entire YAC DNA molecule integrated in the host genome [10]. Also, in the case of the 520-kb YAC, two out of 17 transgenic founder mice identified were also shown to carry the entire YAC DNA molecule, by a systematic PCR analysis throughout the whole locus [12]. In both cases, the high number of DNA-positive transgenic founder mice allowed adequate finding of individuals carrying intact YAC DNA molecules. Furthermore, in the case of the 250-kb YAC molecule, carrying the mouse tyrosinase gene, we could actually rescue the albino phenotype of the host, and hence functionally demonstrate the integrity of the locus included within the genomic transgene. We have coined the name of "ICSI-mediated YAC transfer" or IMYT for our improved technique [10]. With the dissemination of IMYT approaches we believe that the use of large genomic constructs, such as YACs, in standard mouse transgenic experiments can be expanded. In addition, we envisage that this technique can be easily adapted to allow the use of genomic-type transgenes in other species, vertebrates and invertebrates, where pronuclear microinjection has been historically difficult or less efficient.

A novel active method has been developed which combines ICSI with recombinases or transposases to increase transfection efficiency. This technique has been termed "Active Transgenesis", and implies that the transgene is inserted into the host genome by enzymes supplied into the oocyte during transgene introduction. Among approaches utilizing protein recombinases (RecA) [12, 13] or transposases, the hyperactive Tn5 transposase protein (Tn5p) was by far the most efficient method for introducing the transgene in a transposon along with spermatozoa into unfertilized oocytes (TN:ICSI) [14].

MOUSE ICSI-MEDIATED TRANSGENESIS METHOD

ICSI-mediated transgenesis is based on the mouse ICSI protocol described in [15] (Fig. **1**). Briefly, oocyte are obtained from superovulated mice, cumulus cells are dispersed by 5 minutes incubation in M2 medium containing 350 IU/ml hyaluronidase, washed a couple of times in fresh M2 medium and kept at 37°C in KSOM for 30 min before the ICSI procedure.

Epididymal spermatozoa are collected from 8- to 12-week-old mouse males and can be used fresh or freeze-thawed. Live offspring have been reported after ICSI with isolated sperm heads [16], sperm frozen without cryoprotection [17], or even with freeze-dried sperm [18]. Thus, the use of freeze-thawed sperm is advisable because, unlike fresh sperm, freeze-thawed sperm from the same male can be stored and used for several experiments, which reduces the number of sacrificed males. Epididymal sperm can be aliquot in 50 μL vials, frozen in liquid nitrogen, and stored for one month at minus 80 degrees Celcius. Moreover, freezing without cryopreservant produces demembranation of sperm cells and decapitation of spermatozoids. Demembranation is required to allow the DNA to interact with the exposed perinuclear theca of spermatozoa; when using fresh sperm, this has to be achieved chemically by incubation in Triton X-100 [9]. Decapitation is needed for zygote activation and embryo development, and tedious mechanical decapitation by Piezo pulses is required in fresh samples. Direct contact of mouse spermatozoa with liquid nitrogen did not affect their ability to activate injected oocytes, but severely restricted subsequent *in vitro* embryo development to blastocyst stage. Tris-EDTA buffer and M2 were also shown to be better sperm freezing extenders than DPBS, allowing higher developmental potential.

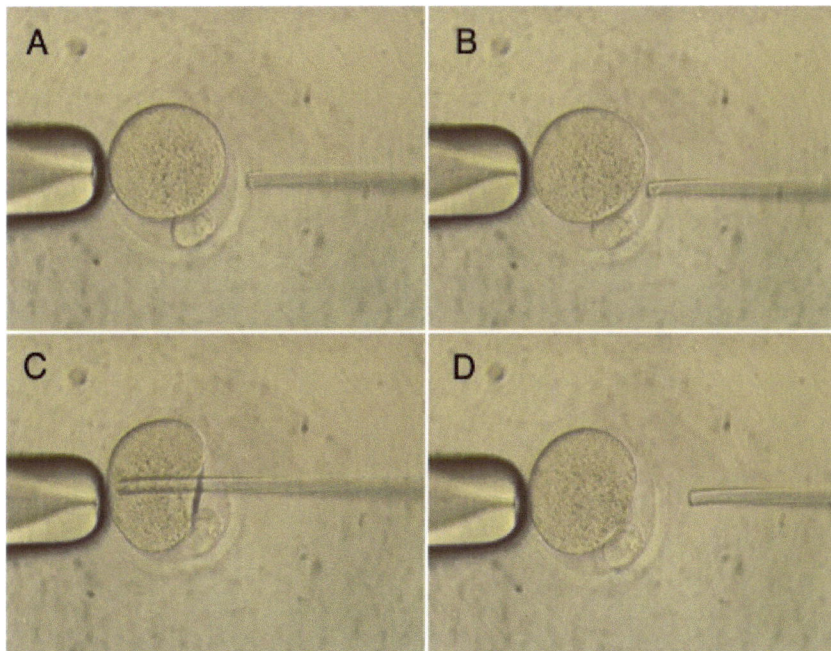

Figure 1: Intracytoplasmic Sperm Injection of a Mouse Sperm Head into a MII Oocyte.

With the help of a piezo unit, the injection pipette has penetrated the zona pellucida of the oocyte. The sperm head, decapitated by the freeze-thawing procedure, after being placed at the tip of the microinjection pipette, is now ready for its deep ooplasmic insertion and deposition. Picture taken on an inverted microscope Nikon TE300 using Hoffman modulation contrast at an amplification of × 400

ICSI can be performed with dead spermatozoa [12]. Freeze-dried or snap-frozen spermatozoa are dead and their plasma membrane disrupted and can not heal, due to a lack of underlying cytoplasm, but their chromosomes remains intact for several months [19]. However, once the spermatozoa are rehydrated or defrosted, their chromosomes start to degenerate. After rehydratation, Na^+ and Ca^{++} rich extracellular medium activates sperm endonucleases that rapidly degrade sperm DNA [20]. For this reason, it is very important to minimize the time interval between the disruption of the sperm plasma membrane and sperm injection into oocytes.

For DNA integration, sperm must be incubated on ice for 2 minutes with an equal volume of transgene construction. Then, one volume of DNA construction: sperm complex is mixed with five volumes of M2-containing 10% polyvinyl-pyrrolidone (PVP) to decrease stickiness. Sperm can be kept at 5°C until use, but must be injected in the next hour. The ICSI dish contains a manipulation drop (M2 medium), a couple of sperm drops (5 volumes M2/10% PVP and 1 volume of M2 medium, as above) where sperm mixture is added, and a 10% PVP in M2 needle-cleaning drop, overlaid with mineral oil to prevent evaporation. The rest of the protocol is the same as with conventional ICSI.

FACTORS AFFECTING THE EFFICIENCY OF ICSI-MEDIATED TRANSGENESIS

Mouse Strain

The use of different mouse strains has been explored allowing extending the method to inbred mouse strains, such as C57BL/6 mice [21], in addition to the albino outbred CD1 [10], hybrid C57CBAF1 [22], or B6D2F1 mice [12, 22], used in our experiments and in previously reported experiments [9]. B6D2F1 mice have been shown to behave optimally for MII-oocytes and sperm cell donors. In the case of hybrids, B6D2F1 sperm cells provided higher embryo development than equivalent cells from C57CBAF1 mice [22].

Effect of DNA Concentration

Besides exploring the use of different buffers [21, 22], the DNA concentration has been assessed as a potential influence on ICSI-mediated transgenesis. Significantly higher plasmid DNA (up to 15 ng/μl) and YAC DNA concentrations (up to 4 ng/μl) could be used with ICSI, as compared with equivalent procedures by pronuclear microinjection [12]. Usually, DNA concentrations exceeding these values are toxic and result in severe arrest during early embryogenesis. In the case of YAC DNA, toxicity can also be enhanced by the presence of tracers of contaminants during the DNA isolation method.

The viability of the embryos/fetuses decreases dramatically when a certain threshold is exceeded. This effect might result from the high efficiency of transgene integration obtained by ICSI-mediated transgenesis, which may lead to chromosomic alterations in microinjected embryos. It has been reported that incubation of spermatozoa with exogenous DNA leads to paternal chromosomal breaks [23, 24]. The frequency of normal karyoplates at blastocysts stage was reported to decrease 2-fold when ICSI-mediated transgenesis was executed with frozen-thawed spermatozoa incubated with 5 ng/μl of pEGFP [24]. These results suggest that incorporation of transgenic DNA into the embryonic genome by ICSI-mediated transgenesis may be directly correlated with paternal chromosome breaks. The relationship between transgene concentration and transgenesis efficiency was similar for the YAC constructs tested. When the YAC construct was used, the efficiency rate of the procedure increased with DNA concentration up to a threshold of 3.6 ng/μl, decreasing when a higher transgene concentration was used. Interestingly, this threshold for the plasmid construct was observed with much higher DNA concentrations. Taking in consideration that the kilobase magnitude of pEGFP is about 100 times smaller than the one of the YAC tested, these observations suggest that for smaller constructs, higher transgene concentrations may be used in order to ensure optimal ICSI-mediated transgenesis results. Moreover, because small plasmids and large YACs require different preparation methods, our results can also reflect a difference in the concentration of co-purified contaminants in YAC and plasmid DNA samples. (See Table **2**, at foot of this Chapter.)

Table 2: ICSI-Mediated Transgenesis Efficiency Obtained for a 510 kb YAC (hAPPy) Construct and a 5.4 kb Plasmid EGFP Transgene Tested at Different Concentrations

Construct (ng/μl)	Injected Oocytes (Sessions)	Surviving oocytes (%)	2-Cell Embryos Transferred (Recipients)	Live Offspring or D14 Fetuses (%)	Transgenic Offspring or D14 Fetuses (%)
YAC (2)	346 (7)	244 (71)	215 (10)	20 (9)	2 (10)[a]
YAC (3.6)	350 (7)	311 (89)	228 (12)	52 (23)	11 (21)[b]
YAC (5)	224 (5)	180 (80)	175 (8)	30 (17)	4 (13)[a]
pEGFP (1)	309 (7)	233 (75)	194 (11)	23 (12)	9 (39)*
pEGFP (6)‡	219 (6)	195 (78)	163 (8)	22 (13)	10 (45)*
pEGFP (15)	313 (6)	255 (81)	230 (14)	30 (13)	21 (70)†
pEGFP (25)	102 (2)	73 (72)	65 (3)	0	-

Gametes for all experiments were collected B6D2 mice.
[a, b and *, †] Values with different superscripts are significantly different (Z-test, $P<0.05$).
‡Data also showed in the reference [2].

The integration rate after ICSI-mediated transgenesis is different from what it was described by Brinster *et al.* for pronuclear microinjection [25], which seems to remain constant when DNA is injected at concentrations above 1 ng/μl. According to our results with pEGFP, the efficiency of the ICSI-mediated transgenesis procedure with plasmid constructs may be significantly improved by increasing the transgene concentration up to values significantly higher than the ones commonly used for pronuclear microinjection. In our study, the efficiency of the ICSI-mediated transgenesis procedure with pEGFP was significantly enhanced by increasing the transgene presence up to concentrations as high as 15 ng/μl [12]. These results suggest that the opportunities for transgene integration might be far superior when the transgene is delivered by ICSI than when it is delivered by pronuclear microinjection. These observations can be explained in light of the fact that, in ICSI-mediated transgenesis, foreign DNA molecules are present during several steps of sperm nucleus transformation into a functional pronucleus, which is concomitant with important chromatin reorganisation events, such as phosphorylation and replacement of sperm specific histones by maternal ones [26]. By contrast, in transgenesis procedures involving pronuclear microinjection, foreign DNA molecules are delivered in a completely formed pronucleus after completion of the majority of sperm nucleus remodelling events. This may also explain the much higher frequency of founder animal mosaicism obtained as the result of late transgene integration (after first mitosis), when the transgene is introduced by pronuclear microinjection than when it is delivered by ICSI. Most ICSI generated founders do not display mosaicism, indicating that the integration of the transgene took place before first embryo division. In addition, the toxicity of the transgene is lower when the transgene is delivered by ICSI than when it is delivered by pronuclear microinjection. These observations can be explained considering that in ICSI-mediated transgenesis the sperm cell-transgene mixture is delivered intracytoplasmically, while during pronuclear microinjection, a volume of transgenic DNA is directly delivered into one of the pronuclei.

SPERM CELL DAMAGE

The regulated damage of sperm heads (by freeze-thaw cycles or exposure to TritonX-100 detergent) was found to be critical to produce transgenic offspring. [8]. The method using freeze-thawed spermatozoa showed severe chromosomal damage [27] and low offspring rates after embryo transfer. Also it can produce long term effects on the offspring [28]. TritonX-100 is not a natural product associated with cells and its residues could be toxic to oocytes. Recently an improved method has been described for the efficient generation of transgenic mice using a simple pretreatment of spermatozoa with 10 mM NaOH

(Table **1**). These spermatozoa lost their plasma membrane and tail, while still maintaining nuclear integrity [29]. This method was found to be suitable for hybrid as well as inbred strains of mouse, such as C57BL/6 and 129X1/Sv. Thus, a simple sperm pretreatment with NaOH before ICSI-Tr resulted in an efficient insertion of an exogenous gene into the host genome. This method allows easy production of transgenic mice, requiring fewer oocytes for micromanipulation than classical methods.

The removal of the acrosome in the mouse is not a prerequisite to produce offspring by ICSI, but it results in earlier onset of oocyte activation and better embryonic development [30]. Recently the plasma membrane and acrosome of the mouse spermatozoa were removed by lysolecithin prior to ICSI [30]. Lysolecithin is a natural cellular hydrolysis product of membrane phospholipids and is unlikely to be toxic to oocytes at residual levels. Two attempts demonstrated that the application of 0.02% lysolecithin-treated spermatozoa to ICSI-mediated transgenesis resulted in high efficiencies of EGFP transgenic mice, 11.4% oocytes injected and 62.5% animals born [31] (Table **1**). These data represent a higher transgenic efficiency for oocytes injected as compared to regular freeze-thawed sperm ICSI-Tr which has an efficiency of 4.6% for oocytes injected.

ACTIVE TRANSGENESIS

The combination of ICSI techniques, using fresh sperm cells, and RecA-coated DNA molecules has also been investigated as a way to enhance integration of the transgene by homologous recombination, and therefore to increase the efficiencies of transgenesis. We compared ICSI-mediated transgenesis using either RecA-free frozen-thawed sperm cells or fresh sperm cells with RecA-coated DNA molecules, and concluded that the efficiencies were similar, and in all cases higher than those obtained with pronuclear microinjection [12]. However, a higher number of mosaic founder mice (30–40%) were observed in the presence of the RecA protein, in contrast to a lower percentage of mosaics detected among the ICSI-mediated transgenic founder obtained with frozen-thawed sperm cells. This suggests that different mechanisms of transgene integration occur in each procedure. In particular, the mosaicism observed with the use of the RecA protein approached the values normally obtained by DNA pronuclear microinjection.

In our study [12], we observed that RecA-complexed and non-complexed rhodamine-labelled enhanced green fluorescent protein (EGFP) DNA attaches differently to fresh sperm cells. We also found that this attachment is different from the one that occurs between RecA-free rhodamine-labelled EGFP DNA and frozen-thawed sperm cells (Fig **2**). When a frozen-thawed sperm sample was co-incubated for either 2 or 120 min with RecA-free rhodamine-labelled EGFP DNA, fluorescent transgene attachment was observed in most sperm cell structures (Fig. **2A**). However, when a fresh sperm sample was incubated for the same period with rhodamine-labelled EGFP, a fluorescent transgene attachment was just detected in the post-acrosomal region of non-motile sperm cells (Fig. **2B**). Finally, when a fresh sperm sample was co-incubated with RecA-complexed rhodamine-labelled EGFP DNA, a generalised fluorescent transgene attachment was observed in head and mid-piece of motile and non-motile sperm cells (Fig. **2C**).

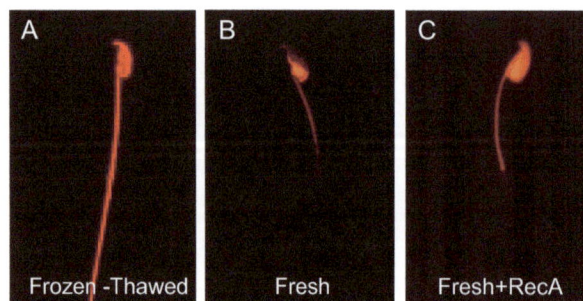

Figure 2: Differences Between Exogenous DNA Binding to Frozen-Thawed and Fresh Mouse Spermatozoa after a 2-Min Co-Incubation Period.

Representative fluorescent images of frozen-thawed mouse spermatozoa co-incubated with rhodamine-labelled DNA (A); fresh spermatozoa co-incubated with rhodamine-labelled DNA (B); and fresh spermatozoa co-incubated with RecA-complexed rhodamine-labelled DNA (C) are depicted at a magnification of ×400. This figure was previously published [12], and is reproduced with permission.

Independently, another group reached comparable transgenic efficiencies using fresh sperm cells and RecA-coated DNA molecules, but did not detect higher frequencies of mosaicism in the corresponding transgenic founder mice produced [13]. Similar to RecA, the use of the hyperactive Tn5 transposase, associated with DNA, has been investigated in ICSI-mediated transgenesis using fresh sperm cells [14]. The results obtained with B6D2F1 mice suggested that the transposase protein greatly enhanced ICSI-mediated transgenesis to values similar to those obtained with frozen sperm cells without associated proteins Table **3**.

Table 3: Comparison of Tested Mouse ICSI-Mediated Transgenesis Procedures with Live and Frozen-Thawed Sperm Cells

	Mouse ICSI-Mediated Transgenesis	
	Immotile Frozen-Thawed Sperm Cells & DNA	**Live Sperm Cells & RecA-Complexed DNA**
DNA Coating	None	RecA-complex
DNA-Sperm Cell Binding	All sperm cell structure	Sperm head and mid-piece
Fertilisation Procedure	ICSI	ICSI
Embryo Production	Low	Medium
Transgenesis Production	Very high	High
Transgene Expression	Frequent	Frequent
Moment of Transgene Integration	Before first mitosis	Before and after first mitosis
Germ line Transmission	Mendelian	Mendelian and mosaic (30-40%)
Efficiency with genomic-type constructs (YAC)	Efficient	Not tested

Although the transgenesis efficiency of the common ICSI-based transgenesis procedure with frozen-thawed sperm cells is similar to that of ICSI with fresh sperm cells and RecA-complexed DNA, the traditional ICSI-mediated transgenesis procedure with frozen-thawed spermatozoa is much more efficient in maintaining a low frequency of founder animal mosaicism. These results suggest that transgene integration takes longer when RecA and fresh sperm cells are used for ICSI-mediated transgenesis, resulting more frequently in delayed transgene integration (after completion of first mitosis). It is known that the regulated damage of sperm heads, by freeze-thaw cycles or exposure to detergents, is critical to improving the production of transgenic offspring with this transgenesis procedure [8]. This improvement comes, possibly, through the facilitation of the sperm cell-transgene interaction after membrane fracture and/or by induction of small sperm DNA breaks that, when combined with the oocyte's DNA repair mechanisms, result in higher transgene integration rates. RecA is a bacterial recombinase that, by binding and coating the ssDNA, has been shown to protect it from shearing and degradation [32]. During previous pig and goat pronuclear microinjection transgenesis attempts, this protective mechanism of RecA has also been suggested to influence the transgene DNA stability [33, 34]. When ICSI-mediated transgenesis is performed with live sperm cells, RecA may facilitate transgene integration, first by protecting the DNA from nucleases present in the sperm solution before sperm injection, and second by protecting it from the intracellular oocyte degradation that usually occurs in its absence (Table **3**). Additionally, RecA may also be interfering with the structural opening of DNA strands that is necessary for the insertion of foreign DNA molecules. However, at present, we cannot discriminate whether the delayed transgene integration observed after ICSI with fresh sperm cells of DNA complexed with RecA is a consequence of the use of the bacterial recombinase, or of the use of intact sperm cells, or both. Future experiments are presently being planned in order to clarify this issue.

CONCLUSIONS

ICSI-mediated transgenesis is the only alternative when very large transgenes of ~500 kb need to be inserted for correctly regulated gene expression. The method enables the generation of genetically modified animals carrying large genomic-type transgenes at a similar efficiency (or even higher) to the values observed for standard-type transgenes in pronuclear microinjection methods. This new technique offers the promising advantage of exporting the gained expertise from the mouse to other mammalian species, enabling the generation of transgenic livestock with higher and reliable efficiencies using large genomic-type constructs. ICSI can be applied to the genetic modification of mouse strains that are difficult to work with (*i.e.* C57BL/6, 129-derived sub-lines). Finally, ICSI-mediated transgenesis can also be viewed as a procedure to ensure efficient germline modification of recipient animals. In mice, the integration of the pronuclear microinjected transgene at a relatively late stage frequently produces a majority of transgenic founder animals that are mosaic. Standard SMGT, using live sperm cells as vehicles for DNA uptake, is also known as a method mostly producing mosaic or non-transmitting transgenic founder animals, a situation that is currently interpreted as the transgenes being maintained and eventually inherited as extrachromosomal structures or episomes. By contrast, mosaicism is less frequent in transgenic founder mice generated by ICSI-mediated transgenesis with frozen-thawed sperm cells. Moreover, in transgenic animals produced by this method, transgene integration in the host genome is the norm, with concomitant transmission of the transgenes to their progeny mostly following the expected Mendelian ratio.

REFERENCES

[1]   Brackett BG, Baranska W, Sawicki W, Koprowski H. Uptake of heterologous genome by mammalian spermatozoa and its transfer to ova through fertilisation. Proc Natl Acad Sci USA 1971; 68: 353-57.

[2]   Lavitrano M, Camaioni A, Fazio VM, Dolci S, Farace MG, Spadafora C. Sperm cells as vectors for introducing foreign DNA into eggs: genetic transformation of mice. Cell 1989; 57: 717-23.

[3]   Brinster RL, Sandgren EP, Behringer RR, Palmiter RD. No simple solution for making transgenic mice. Cell 1989; 59: 239-41.

[4]   Gandolfi F. Sperm-mediated transgenesis. Theriogenology 2000; 53: 127-37.

[5]   Wall RJ. New gene transfer methods. Theriogenology 2002; 57: 189-01.

[6]   Celebi C, Guillaudeux T, Auvray P, Vallet-Erdtmann V, Jegou B. The making of "transgenic spermatozoa". Biol Reprod 2003; 68: 1477-83.

[7]   Kimura Y, Yanagimachi R. Intracytoplasmic sperm injection in the mouse. Biol Reprod 1995; 52: 709-20.

[8]   Perry AC, Wakayama T, Kishikawa H, *et al.* Mammalian transgenesis by intracytoplasmic sperm injection. Science 1999; 284: 1180-83.

[9]   Perry AC, Rothman A, de las Heras JI, *et al.* Efficient metaphase II transgenesis with different transgene archetypes. Nat Biotechnol 2001; 19: 1071-73.

[10]   Moreira PN, Giraldo P, Cozar P, *et al.* Efficient generation of transgenic mice with intact yeast artificial chromosomes by intracytoplasmic sperm injection. Biol Reprod 2004; 71: 1943-47.

[11]   Giraldo P, Montoliu L. Size matters: use of YACs, BACs and PACs in transgenic animals. Transgenic Res 2001; 10: 83-03.

[12]   Moreira PN, Perez-Crespo M, Ramirez MA, *et al.* Effect of transgene concentration, flanking matrix attachment regions, and RecA-coating on the efficiency of mouse transgenesis mediated by intracytoplasmic sperm injection. Biol Reprod 2007; 76: 336-43.

[13]   Kaneko T, Moisyadi S, Suganuma R, *et al.* Recombinase-mediated mouse transgenesis by intracytoplasmic sperm injection. Theriogenology 2005; 64: 1704-15.

[14]   Suganuma R, Pelczar P, Spetz JF, *et al.* Tn5 transposase-mediated mouse transgenesis. Biol Reprod 2005; 73: 1157-63.

[15]   Yoshida N, Perry AC. Piezo-actuated mouse intracytoplasmic sperm injection (ICSI). Nat Protoc 2007; 2: 296-304.

[16]   Kuretake S, Kimura Y, Hoshi K, Yanagimachi R. Fertilisation and development of mouse oocytes injected with isolated sperm heads. Biol Reprod 1996; 55: 789-95.

[17]   Wakayama T, Whittingham DG, Yanagimachi R. Production of normal offspring from mouse oocytes injected with spermatozoa cryopreserved with or without cryoprotection. J Reprod Fertil 1998; 112: 11-17.

[18]   Wakayama T, Yanagimachi R. Development of normal mice from oocytes injected with freeze-dried spermatozoa. Nat Biotechnol 1998; 16: 639-41.

[19]   Ward MA, Kaneko T, Kusakabe H, *et al.* Long-term preservation of mouse spermatozoa after freeze-drying and freezing without cryoprotection. Biol Reprod 2003; 69: 2100-08.

[20]   Tateno H, Kimura Y, Yanagimachi R. Sonication *per se* is not as deleterious to sperm chromosomes as previously inferred. Biol Reprod 2000; 63: 341-46.

[21]   Osada T, Toyoda A, Moisyadi S, *et al.* Production of inbred and hybrid transgenic mice carrying large (> 200 kb) foreign DNA fragments by intracytoplasmic sperm injection. Mol Reprod Dev 2005; 72: 329-35.

[22]   Moreira PN, Jimenez A, Fernandez R, *et al.* Mouse ICSI with frozen-thawed sperm: the impact of sperm freezing procedure and sperm donor strain. Mol Reprod Dev 2003; 66: 98-03.

[23]   Ward MA, Ward WS. A model for the function of sperm DNA degradation. Reprod Fertil Dev 2004; 16: 547-54.

[24]   Szczygiel MA, Moisyadi S, Ward WS. Expression of foreign DNA is associated with paternal chromosome degradation in intracytoplasmic sperm injection -mediated transgenesis in the mouse. Biol Reprod 2003; 68: 1903-910.

[25]   Brinster RL, Chen HY, Trumbauer ME, Yagle MK, Palmiter RD. Factors affecting the efficiency of introducing foreign DNA into mice by microinjecting eggs. Proc Natl Acad Sci USA 1985; 82: 4438-442.

[26]   Green GR, Poccia DL. Phosphorylation of sea urchin sperm H1 and H2B histones precedes chromatin decondensation and H1 exchange during pronuclear formation. Dev Biol 1985; 108: 235-45.

[27]   Yamagata K, Suetsugu R, Wakayama T. Assessment of chromosomal integrity using a novel live-cell imaging technique in mouse embryos produced by intracytoplasmic sperm injection. Hum Reprod 2009; 24: 2490-499.

[28]   Fernandez-Gonzalez R, Moreira PN, Perez-Crespo M, *et al.* Long-term effects of mouse intracytoplasmic sperm injection with DNA-fragmented sperm on health and behavior of adult offspring. Biol Reprod 2008; 78: 761-72.

[29]   Li C, Mizutani E, Ono T, Wakayama T. An efficient method for generating transgenic mice using NaOH-treated spermatozoa. Biol Reprod. 2010; 82: 331-40.

[30]   Morozumi K, Shikano T, Miyazaki S, Yanagimachi R. Simultaneous removal of sperm plasma membrane and acrosome before intracytoplasmic sperm injection improves oocyte activation/embryonic development. Proc Natl Acad Sci USA 2006; 103: 17661-7666.

[31]   Moisyadi S, Kaminski JM, Yanagimachi R. Use of intracytoplasmic sperm injection (ICSI) to generate transgenic animals. Comp Immunol Microbiol Infect Dis 2009; 32: 47-60.

[32]   Chow SA, Honigberg SM, Bainton RJ, Radding CM. Patterns of nuclease protection during strand exchange. recA protein forms heteroduplex DNA by binding to strands of the same polarity. J Biol Chem 1986; 261: 6961-971.

[33]   Maga EA. The use of recombinase proteins to generate transgenic large animals. Cloning Stem Cells 2001; 3: 233-41.

[34]   Maga EA, Sargent RG, Zeng H, *et al.* Increased efficiency of transgenic livestock production. Transgenic Res 2003; 12: 485-96.

[35]   Nakanishi T, Kuroiwa A, Yamada S, *et al.* FISH analysis of 142 EGFP transgene integration sites into the mouse genome. Genomics 2002; 80: 564-74.

CHAPTER 11

Nanobiotechnology and SMGT: Future Perspectives

Vinicius Farias Campos, Fabiana Kömmling Seixas, Odir A. Dellagostin, João Carlos Deschamps and Tiago Collares*

Biotechnology Centre, Federal University of Pelotas, Brazil

Abstract: Nanoparticles are being incorporated into many products of daily use, *e.g.* fillers, pacifiers, catalysts, pharmaceuticals, lubricants, cosmetics, pharmaceuticals, electronic devices or other domestic appliances. These approaches demonstrate the high potential of nanocomposites in cellular biotechnology as well as for the development of more efficient methods to introduce foreign DNA to sperm cells and consequently improve the techniques for transgenic animal technology. This chapter will address the main approaches for the use of nanocomposites for gene delivery and the potential uses in nanoparticle sperm-mediated gene transfer (Nano-SMGT).

Keywords: Nanostructures, Nanotubes, Nanopolymers, Nanoparticles, SMGT, Gene therapy, DNA/RNA delivery, Sperm, Transfection, Transgenesis, Knockdown.

NANOTECHNOLOGY FOR CELLULAR BIOTECHNOLOGY

Nanoscience and nanotechnology are dynamically developing fields of scientific interest in the entire world and have already become key research and development priorities in Europe and North America [1]. Nanotechnology is often regarded as an 'enabling technology'. It exploits properties and phenomena developed at the nanoscale. 'Nano' derives from the Greek word nanos meaning a dwarf. Invention of the first tool for manipulation of atomic structures, a scanning tunneling microscope (STM), can be considered one of the most important milestones in the development of nanotechnology. Since the 1990s there has been a very rapid increase in the implementation of nanotechnologies. Estimates are that, by 2014, more than 15% of all products on the global market will have some kind of nanotechnology incorporated into their manufacturing process [2].

Currently we have two main classes of nanotechnology products: nanomaterials fixed on a substrate and free nanoparticles. Nanosilver, various forms of carbon, zinc oxide, titanium dioxide and iron oxide make up the majority of nanomaterials in use. Nanoparticles are being incorporated into many products of daily use, *e.g.* fillers, pacifiers, catalysts, pharmaceuticals, lubricants, cosmetics, pharmaceuticals, electronic devices or other domestic appliances [3].

Achievements in this area find practical applications in many fields of industry and daily life, *e.g.* in medicine – from therapeutic and diagnostic medicine to surgery. Utilisation of nanoparticles in medical procedures has also been developing rapidly, for example as imaging tools, phototherapy agents and as gene/drug delivery carriers [4].

Magnetic nanoparticles and functionalised nanotubes are already used in diagnostics and molecular biology [5]. One can predict that nanoparticles with magnetic properties, such as magnetite and ferrofluidic nanoparticles, will soon find application in transport of chemical factors [6], *e.g.* in directed drug delivery in cancer therapy [7] and in biomanipulation, as DNA or RNA transfection [8].

However, in order to make major progress for the applications described above, it is of paramount importance to improve transfection efficiencies of nanocomposite vectors. Unfortunately, the transfection efficiency of a non-viral gene vector is affected by many factors such as nuclease digestion, cellular uptake,

---

*Address correspondence to Tiago Collares: Biotechnology Centre, Federal University of Pelotas, University Campus, CEP 96010-900, Postal Code 354, Pelotas, RS – Brazil; Tel: +555332757588; E-mail: collares.t@gmail.com

endosomal escape, intracellular stability, nuclear transfer and intra-nuclear transcription. Therefore, to deliver transgenes into the nucleus of cells efficiently it is necessary to equip this non-viral vector with various functional elements. Such elements may include: a component for cell-specific delivery; an element for endosomal escape; and a nuclear localisation signal (NLS) for enhancing nuclear delivery [9]. Addition of these elements may permit the transgene to express at the appropriate time and location. Several advances in gene delivery technology have been developed in the area of gene therapy and these studies could be used for gene transfection in several cell types. Some commercial nanoparticles are already available and promise to improve common methods of cell transfection. Recently, potential uses of nanocomposites in spermatozoa have been investigated [10], as well as for sperm mediated gene transfer [11]. Also, the toxic effects of these nanoparticules have been studied in human spermatozoa [12].

These approaches demonstrate the high potential of nanocomposites in cellular biotechnology as well as for the development of more efficient methods to introduce foreign DNA to sperm cells and consequently improve the techniques for transgenic animal technology. This chapter will address the main approaches for the use of nanocomposites for gene delivery and the potential uses in nanoparticle sperm-mediated gene transfer (Nano-SMGT).

MOLECULAR DELIVERY INTO SPERMATOZOA USING NANOCOMPOSITES

Several methods have been described to introduce DNA vectors into sperm cells, such as pure plasmids, electroporation, receptor-mediated gene transfer, particle guns, viral vectors or lipofection [13-15]. Each system has benefits and limitations, and to date there is no ideal method for gene transfer. Both viral and nonviral systems have been used for gene transfection, and each process has unique characteristics. Viral-mediated gene transfer has a number of advantages over nonviral methods, such as higher DNA uptake and more efficient gene expression. However, vector inactivation, development of replication-competent virions, and need for a relatively large-scale infrastructure for their production represent serious disadvantages, as does immunogenicity and potential oncogenic properties [16], at least where *in vivo* applications are envisaged (such as *in vivo* forms of SMGT). Such drawbacks would be of particular concern were SMGT ever to be considered for human use. In contrast, cationic liposome/DNA complexes for gene transfer (lipofection) into sperm cells have frequently been used as nonviral DNA carriers for gene transfection [17]. However, there are numerous issues that must be resolved. These include the finite lifetime of the transfection complex, inactivation of the gene/liposome complex, degradation in the endosomal compartment, and transfer to the nucleus. The majority of nanoparticle research has been carried out by material scientists, but recent trends have brought these tools into the hands of biologists.

Nanoparticles have found broad niches in biology. Since the late 1970s, nanoparticles have been used for drug delivery [18]. Nanoparticle-mediated gene delivery has recently emerged as a promising tool for gene therapy strategies [19]. Several nanocomposites could be used for gene delivery into spermatozoa. For this potential to be accomplished, an ability to target the nanoparticles to specific sites through surface chemistry would be important. High specificity can be introduced by using biological moieties that process lock-and key interactions, including those observed in antibody-antigen and enzyme-substrate recognition. In surface modification with biomolecules the surface character is critically important, as is the ability for the immobilised molecules to retain their native conformation and binding profile. Owing to their magnetic properties, biocompatibility, and physical properties such as size, shape, and surface characteristics, iron oxide nanoparticles have been recognised as a promising tool for the site-specific delivery of drugs and other bioactive molecules including DNA vectors.

MAGNETOFECTION FOR NANO-SMGT

The term nanoparticle (NP) refers to particles less than 1 μm in size, typically less than 200 nm. Thus they are of a similar length scale to biomolecules, making NPs ideal for combining with biomolecules for medical applications [19]. Due to their size, they can penetrate practically any tissue, including the blood brain barrier, and tissues protected by tight junctions. In addition, the resultant increase in surface area to volume ratio compared to other therapeutic molecules creates more scope for surface modifications and conjugation to functionally specific ligands [20]. However the cell and nuclear membranes present a substantial hurdle to the use of magnetic NPs that are not actively translocated into cells [21]. Currently several methods are been adopted to overcome this, each with its own limitations.

Magnetic nanoparticles are being increasingly used in a number of biological and medical applications, including cell sorting and transfection [22]. Magnetofection is a new method for gene transfer that involves the use of magnetic force and plasmid DNA (pDNA)/magnetic bead complexes, and it has been developed for enhancing delivery of gene vectors to target cells [23]. For magnetofection, pDNA was interacted with magnetic beads and attracted to target cells by magnetic force in order to accumulate on the cell surface.

This method associates nucleic acids or other vectors with magnetic nanoparticles coated with cationic molecules. Generally, the magnetic nanoparticles are made of iron oxide, which is fully biodegradable, coated with specific cationic proprietary molecules varying upon applications. Their association with the gene vectors is achieved by salt-induced colloidal aggregation and electrostatic interaction. The resulting molecular complexes are then targeted to and endocytosed by cells, supported by an appropriate magnetic field. Membrane architecture and structure stay intact in contrast to other physical transfection methods that damage, create hole or electroshock the cell membranes. In addition the magnetic nanobeads are efficiently cleared from the cells and tissues, and are not toxic at the recommended doses and even higher.

The association of magnetic nanoparticles and several other compounds to improve transfection efficiency has been reported in several studies. Ino *et al.* [24] demonstrated the production of magnetic cationic liposomes (MCLs), containing 10-nm magnetite nanoparticles in order to improve the accumulation of magnetite nanoparticles in target cells through electrostatic interactions between MCLs (positively charged) and the cell membrane (negatively charged). Since MCLs are positively charged, pDNA (negatively charged) interacts with the MCLs electrostatically. This method increases significantly the DNA uptake by cultured cells, however cellular damages were increased. Another magnetofection variation uses self-assembled ternary complexes of cationic magnetic nanoparticles, plasmid DNA and cell-penetrating Tat peptide. Tat is a highly cationic peptide derived from HIV-1 tat protein, with a linear sequence of 13 amino acids. This specific sequence carries a transmembrane signal and a nuclear localisation signal. The membrane absorption is primarily promoted by ionic interaction between the cationic charges of the Tat peptide and anionic charges of the phospholipid heads in biomembranes. Benefiting from its cell membrane penetrating property, the Tat peptide is capable of mediating intracellular delivery of many different types of cargos [25], including magnetic nanoparticles [26]. While both magnetic field-mediated gene transfer and Tat peptide-assisted delivery offer attractive features, the combination of the two methods has been documented to provide additional advantages in improving targeted transgene expression [26].

Magnetofection for gene delivery in sperm cells was recently demonstrated by Kim *et al.* [11], producing transgenic boar embryos expressing enhanced green fluorescent protein. The complex formed between plasmid DNA and MNPs was bound on ejaculated boar spermatozoa at a higher efficiency compared to methods using DNA alone or lipofection. This study demonstrates the great potential of nanotechnology for Nano-SMGT and consequently for the improvement in the production of transgenic animals.

NANOTUBES FOR NANO-SMGT

Carbon nanotubes (CNTs) were described in 1991 by Iijima [27]. There are two classes of CNTs: single-walled nanotubes (SWCNTs), which consist of a single graphite sheet seamlessly wrapped into a cylindrical tube with a diameter between 0.4 and 2.5 nm, and multiwalled carbon nanotubes (MWCNTs), which comprise more layers of graphite sheet with different diameters of up to 100 nm. The length of the tubes ranges from a few nanometers to a few micrometers. Their unique structure offers CNTs with excellent physical and chemical properties [28] that enable wide industrial applications.

In the past decade, we have witnessed the rapid development of nanotechnology in many fields. For example, the applications of CNTs in medicine have been highlighted in several review papers with a focus on cancer treatment.

Vertically aligned carbon nanotubes grown by plasma-enhanced chemical vapor deposition (PECVD) [29] have ferromagnetic catalyst nickel particles enclosed in their tips. This structure makes nanotubes respond to magnetic agitation. The momentum of the carbon nanotubes can be used to penetrate cell membranes (nanotube spearing) and thereby shuttle macromolecules immobilised on the carbon nanotubes into cells. This should be more controllable than other nanocomposite-based molecular delivery approaches [30] because the magnetic field strength, the nanotube speed and the period of spearing can be used to tune the penetration efficiency. Indeed, in previous research, no cytotoxicity was observed at low dose concentration (< 10 µM), which implies that carbon nanotubes are biocompatible [31].

Several studies have demonstrated that DNase activity present in the seminal plasma and spermatozoa [32] has a negative effect in SMGT protocols [15]. The present knowledge shows that carbon nanotubes protect DNA strands against nucleases during cellular delivery [33]. Specifically, when bound to single-walled carbon nanotubes, DNA vector are protected from enzymatic cleavage and interference from nucleic acid binding proteins. Recently, it was proposed that physical properties of carbon nanotubes can influence their internalisation by living cells, mainly in relation to nanotube length [34]. Considering that spermatozoa of each species have a particular membrane structure and intracellular environment, which requires the development of specific SMGT protocols, carbon nanotubes seem to be an interesting alternative for transfecting exogenous DNA into spermatozoon nuclei.

CONCLUSIONS

In recent years nanotechnology has reached a stage that allows nanoparticles with magnetic properties to be functionalised with specific ligands for various clinical applications, such as site-specific drug/gene delivery, magnetic resonance imaging, hyperthermic treatment for cancer cells and tumor targeting. However, studies using nanoparticles for transfection of sperm cells are still in an early stage. If the potential of nanobiotechnology is to deliver products to cells, why not targeting for sperm cells? We believe that within a few years, Nano-SMGT will be developed into an efficient mechanism for obtaining transgenic animals.

REFERENCES

[1]   Fernandez MPA, Hullmann A. A boost for safer nanotechnology. Nano Today 2007; 2: 56.

[2]   Dawson NG. Sweating the small stuff: environmental risk and nanotechnology. Bioscience 2008; 58: 690.

[3]   Nel A, Xia T, Madler L, Li N. Toxic potential of materials at the nanolevel. Science 2006 3; 311: 622-27.

[4]   Verma A, Uzun O, Hu Y, Hu Y, Han HS, Watson N, *et al.* Surface-structure-regulated cell-membrane penetration by monolayer-protected nanoparticles. Nat Mater 2008; 7: 588-95.

[5]   Ngomsik AF, Bee A, Draye M, Cote G, Cabuil V. Magnetic nano- and microparticles for metal removal and environmental applications: a review. Comptes Rendus Chimie 2005; 8: 963-70.

[6]   Liu WT. Nanoparticles and their biological and environmental applications. J Biosci Bioeng 2006; 102: 1-7.

[7]   Arruebo M, Fernandez-Pacheco R, Ibarra MR, Santamaria J. Magnetic nanoparticles for drug delivery. Nano Today 2007; 2: 22-32.

[8]   Zohra FT, Chowdhury EH, Akaike T. High performance mRNA transfection through carbonate apatite-cationic liposome conjugates. Biomaterials 2009; 30: 4006-13.

[9]   Moriguchi R, Kogure K, Akita H, Futaki S, Miyagishi M, Taira K, *et al.* A multifunctional envelope-type nano device for novel gene delivery of siRNA plasmids. Int J Pharm 2005; 301: 277-85.

[10]   Makhluf SB, Abu-Mukh R, Rubinstein S, Breitbart H, Gedanken A. Modified PVA-Fe3O4 nanoparticles as protein carriers into sperm cells. Small 2008; 4: 1453-58.

[11]   Kim TS, Lee SH, Gang GT, Lee YS, Kim SU, Koo DB, *et al.* Exogenous DNA Uptake of Boar Spermatozoa by a Magnetic Nanoparticle Vector System. Reprod Domest Anim 2009.

[12]   Wiwanitkit V, Sereemaspun A, Rojanathanes R. Effect of gold nanoparticles on spermatozoa: the first world report. Fertil Steril 2009; 91: e7-e8.

[13]   Yang SY, Sun JS, Liu CH, Tsuang YH, Chen LT, Hong CY, *et al. Ex vivo* magnetofection with magnetic nanoparticles: a novel platform for nonviral tissue engineering. Artif Organs 2008; 32: 195-04.

[14] Hoelker M, Mekchay S, Schneider H, Bracket BG, Tesfaye D, Jennen D, *et al.* Quantification of DNA binding, uptake, transmission and expression in bovine sperm mediated gene transfer by RT-PCR: effect of transfection reagent and DNA architecture. Theriogenology 2007; 67: 1097-07.

[15] Collares T, Campos VF, Seixas FK, Cavalcanti PV, Dellagostin OA, Moreira HL, *et al.* Transgene transmission in South American catfish (Rhamdia quelen) larvae by sperm-mediated gene transfer. J Biosci 2010; 35: 39-47.

[16] Lehrman S. Virus treatment questioned after gene therapy death. Nature 1999; 401: 517-18.

[17] Harel-Markowitz E, Gurevich M, Shore LS, Katz A, Stram Y, Shemesh M. Use of sperm plasmid DNA lipofection combined with REMI (restriction enzyme-mediated insertion) for production of transgenic chickens expressing eGFP (enhanced green fluorescent protein) or human follicle-stimulating hormone. Biol Reprod 2009; 80: 1046-52.

[18] Kreuter J, Tauber U, Illi V. Distribution and elimination of poly(methyl-2-14C-methacrylate) nanoparticle radioactivity after injection in rats and mice. J Pharm Sci 1979; 68: 1443-47.

[19] de la Fuente JM, Berry CC, Riehle MO, Curtis AS. Nanoparticle targeting at cells. Langmuir 2006; 22: 3286-93.

[20] Sajja HK, East MP, Mao H, Wang YA, Nie S, Yang L. Development of multifunctional nanoparticles for targeted drug delivery and noninvasive imaging of therapeutic effect. Curr Drug Discov Technol 2009; 6: 43-51.

[21] Smith CA, de la Fuente J, Pelaz B, Furlani EP, Mullin M, Berry CC. The effect of static magnetic fields and tat peptides on cellular and nuclear uptake of magnetic nanoparticles. Biomaterials 2010; 31: 4392-00.

[22] Kadota S, Kanayama T, Miyajima N, Takeuchi K, Nagata K. Enhancing of measles virus infection by magnetofection. J Virol Methods 2005; 128: 61-66.

[23] Gersting SW, Schillinger U, Lausier J, Nicklaus P, Rudolph C, Plank C, *et al.* Gene delivery to respiratory epithelial cells by magnetofection. J Gene Med 2004; 6: 913-22.

[24] Ino K, Kawasumi T, Ito A, Honda H. Plasmid DNA transfection using magnetite cationic liposomes for construction of multilayered gene-engineered cell sheet. Biotechnol Bioeng 2008; 100: 168-76.

[25] Berry CC. Intracellular delivery of nanoparticles *via* the HIV-1 tat peptide. Nanomedicine (Lond) 2008; 3: 357-65.

[26] Song HP, Yang JY, Lo SL, Wang Y, Fan WM, Tang XS, *et al.* Gene transfer using self-assembled ternary complexes of cationic magnetic nanoparticles, plasmid DNA and cell-penetrating Tat peptide. Biomaterials 2010; 31: 769-78.

[27] Iijima S. Helical microtubules of graphitic carbon. Nature 1991; 354: 56-58.

[28] Niyogi S, Hamon MA, Hu H, Zhao B, Bhowmik P, Sen R, *et al.* Chemistry of single-walled carbon nanotubes. Acc Chem Res 2002; 35: 1105-13.

[29] Ren ZF, Huang ZP, Xu JW, Wang JH, Bush P, Siegal MP, *et al.* Synthesis of large arrays of well-aligned carbon nanotubes on glass. Science 1998; 282: 1105-07.

[30] Pantarotto D, Briand JP, Prato M, Bianco A. Translocation of bioactive peptides across cell membranes by carbon nanotubes. Chem Commun (Camb) 2004; 7: 16-17.

[31] Hu H, Ni YC, Montana V, Haddon RC, Parpura V. Chemically functionalised carbon nanotubes as substrates for neuronal growth. Nano Letters 2004; 4: 507-11.

[32] Lanes CF, Sampaio LA, Marins LF. Evaluation of DNase activity in seminal plasma and uptake of exogenous DNA by spermatozoa of the Brazilian flounder Paralichthys orbignyanus. Theriogenology 2009; 71: 525-33.

[33] Wu Y, Phillips JA, Liu H, Yang R, Tan W. Carbon nanotubes protect DNA strands during cellular delivery. ACS Nano 2008; 2: 2023-28.

[34] Raffa V, Ciofani G, Nitodas S, Karachalios T, D'Alessandro D, Masini M, *et al.* Can the properties of carbon nanotubes influence their internalisation by living cells? Carbon 2008; 46: 1600-10.

CHAPTER 12

Models of Transgene Integration and Transmission

Ilaria Sciamanna and Corrado Spadafora[*]

Italian National Health Institute, Viale Regina Elena 299, 00161 Rome, Italy

Abstract: The discovery that spermatozoa of essentially all species can spontaneously take up exogenous DNA and deliver it to oocytes at fertilization suggested their use as vectors of foreign information for the generation of transgenic animals. Hence a variety of protocols of sperm-mediated gene transfer (SMGT) have been developed in numerous species. The outcomes in stability and transmission of the exogenous sequences are highly heterogeneous, casting doubt on whether the generated animals are truly transgenic and leaving an open question as to the final fate of the DNA molecules delivered by sperm cells. Two findings have contributed to clarify the underlying molecular mechanism of SMGT: i) the discovery that DNA-loaded demembranated spermatozoa used in ICSI assays increase the yield of genuine transgenic mice, and ii) the identification of a reverse transcriptase (RT) activity in mature spermatozoa. When membrane-disrupted sperm cells are incubated with DNA prior to microinjection in oocytes, the foreign sequences are integrated in the host genome and a high proportion of the offspring are genuinely transgenic. In intact spermatozoa, instead, the foreign DNA binds to the plasma membrane, hence is internalised and undergoes a two-step reaction, first of transcription in RNA and then of reverse-transcription in cDNA copies; these cDNAs behave as transcriptionally competent retrogenes, are propagated as non-integrated extrachromosomal structures and are transmitted to the progeny in a non-Mendelian fashion. We have called this phenomenon sperm-mediated "reverse" gene transfer (SMRGT). These results point out the central role of the plasma membrane in the final fate of exogenous sequences as either integrated transgenes or extrachromosomal retrogenes. The latter underscore a previously unrecognised transgenerational genetics, and a form of non-Mendelian inheritance, mediated by an RT-dependent mechanism.

Keywords: SMGT, Sperm, DNA binding, Sperm plasma membrane, Demembranated, Sperm cells, ICSI, Reverse Transcriptase, Episomal DNA, Retro-Genes, SMRGT, Non-Mendelian Genetics.

INTRODUCTION

This chapter focuses on the molecular mechanism underlying the genesis, propagation and expression of novel genetic and phenotypic traits produced *via* sperm-mediated gene transfer (SMGT). As discussed elsewhere in this volume, SMGT was originally reported as a protocol in which intact spermatozoa could take up foreign DNA and deliver it to oocytes during fertilization [1]. In that framework, spermatozoa were viewed as "neutral" carriers of exogenous sequences. Subsequent studies have however subverted this view and disclosed a complex network of functions in sperm cells, which produce new traits in the offspring and entail a non-conventional mode of genetic transmission.

As recalled in more depth in chapters 1 and 2, SMGT initially raised skepticism in the scientific community as a reliable method for producing transgenic animals, mainly due to the low reproducibility emerging from attempts in various laboratories to develop SMGT protocols in murine models [2] and in other species [3-5]. The ambiguity concerned important aspects of the method, both quantitative and qualitative, which contrasted with the more established microinjection methods; as a matter of fact, the differences between SMGT- and microinjection-generated animals are numerous and more relevant than their similarities. Our own early studies pointed out a great discontinuity among results – a state of affairs that has become amenable to rationalisation only recently. Some points were particularly difficult to rationalise. First, the proportion of SMGT-positive mice that clearly harboured the DNA sequences originally incubated with spermatozoa was extremely variable, ranging from 0% in some experiments to

---

*Address correspondence to Corrado Spadafora: Italian National Health Institute, Viale Regina Elena 299, 00161 Rome, Italy; Tel: 39 (0)6 49903117; E-mail: corrado.spadafora@iss.it

85% in others [6]; such high variability has little in common with the rather consistent rate of transgenesis obtained by microinjection. Second, foreign sequences generated and propagated *via* SMGT exhibited a highly heterogeneous distribution in offspring, during development and throughout the adult tissues, both somatic and germline, with highly variable expression, suggesting that qualitative differences are intrinsic features of SMGT-generated individuals. Over the years, evidence from our and other laboratories accumulated in apparently contradictory directions: in some SMGT experiments exogenous sequences were only detectable in embryos, while in others they were propagated to adult animals, sometimes integrating in the host genomes or more often maintained as extrachromosomal structures, variably distributed in differentiated tissues, transcriptionally competent or silent, transmitted to the F1 progeny in a non-Mendelian fashion or limited to F0 founders. On the whole, this heterogeneous body of results suggests that the process triggered by foreign DNA binding to intact spermatozoa shares little analogy with genuine transgenesis, raising crucial questions on the real nature of the newly generated traits.

By contrast, transgenic mice generated by DNA microinjection in zygotic nuclei are characterised by stable features: they are obtained with a reproducible rate of 10-15%, harbour transgenes that are stably and homogeneously integrated in the genome of both somatic and germline cells, transmitted as Mendelian characters to the progeny and usually expressed in specific manners depending on the promoter specificity. In addition to transgenic mice, transgenics in a variety of species, including farm animals of commercial interest, also respond to the same genetic rules.

These different genetic features suggest that conventional and SMGT transgenesis activate different modes of genetic transmission. Indeed, we now know that mature spermatozoa are endowed with an endogenous reverse transcriptase (RT) activity [7], a crucial component activated by the uptake of foreign RNA [8] or DNA [9] sequences in sperm cells and capable of reverse-transcribing them in cDNA copies. RT is emerging as a novel source of genetic and phenotypic traits unlinked to chromosomal genes, with a key role in SMGT process that can account for its apparently heterogeneous outcomes.

SMGT AS A REVERSE TRANSCRIPTASE-DEPENDENT PHENOMENON: THE GENESIS OF RETRO-GENES

Understanding the sperm RT activity is key to clarify the real nature of the SMGT process and disclose the molecular mechanism through which SMGT-dependent traits are generated in the offspring. Fig. **1** schematises the SMGT process and highlights some essential aspects of the underlying molecular mechanisms. Briefly: the binding of exogenous DNA or RNA molecules with intact spermatozoa and their internalisation into sperm nuclei are mediated by specific protein factors on the sperm cell membrane (see Chapter 2 for in-depth description); this triggers an RT-dependent mechanism that reverse transcribes foreign sequences of both classes into cDNA copies. Exogenous RNA is directly reverse-transcribed in a one-step reaction, whereas exogenous DNA needs a two-steps reaction, being first transcribed in RNA by an RNA polymerase present in spermatozoa [10] and then reverse-transcribed in cDNA copies. Only a small proportion of the newly synthesized retrocopies are retained in the sperm nucleus; the majority are released as extrachromosomal structures from the sperm cell in the medium and therein are taken up by other spermatozoa suspended in the medium. Within 60 min the majority of the sperm population (>80%) is loaded with both the original exogenous sequences and their reverse-transcribed copies. When these spermatozoa are used in *in vitro* fertilization (IVF) assays, both DNA populations are delivered to oocytes and propagated throughout preimplantation development. At later developmental stages (fetuses), and in tissues of young individuals, the reverse-transcribed population becomes predominant, if not exclusive, compared to the original DNA, as determined by direct PCR amplification [9]. Events of integration of the reverse-transcribed sequences in the sperm genome can occur, but they represent rare exceptions [11]. Sequence analysis indicates that integrations occur in one, or very few, preferential site(s) located in particular nucleosomal domains, *i.e.* chromatin portions in which histones have not been replaced by protamines during spermiogenesis. According to a well-accepted model, nucleohistone chromatin components are placed between adjacent protamine domains in close contact with the nuclear scaffold [12], as shown in Fig. **1**. Sperm nucleohistone domains are sensitive to micrococcal nuclease digestion [13] and, in contrast with the tightly packaged and virtually inaccessible protamine domains, they represent chromatin stretches accessible to foreign molecules and potentially favourable sites for exogenous DNA integration.

We therefore focussed on SMGT as an RT-dependent source of novel molecules behaving as transcriptionally competent, non-integrated retrogenes able to generate novel phenotypic traits in animals. We have called this phenomenon Sperm-Mediated "Reverse" Gene Transfer (SMRGT). Paradoxically, no coding gene that would account for the expression of new phenotypic traits can be detected by Southern analysis, but only by PCR amplification [9], supporting the conclusion that coding retrogenes are present in low copy number (fewer than one copy per genome), below the resolution power of Southern analysis.

These features, *i.e.* extrachromosomal, low-copy number and mosaic distribution among organs and within organs, suggest that retrosequences can only replicate in a restricted subpopulation of cells. Furthermore, this leads to the hypothesis that the bulk of differentiated cells in tissues may not be permissive for replication of retrosequences, whereas less differentiated stem or progenitor cells may offer a more permissive environment, which would explain the markedly favoured persistence of these sequences throughout embryonic life compared to adult tissues, where they decline over time. By analogy, this recalls the concept of stem cell 'niche' as a small cell population retaining 'stemness' features that distinguish them from the differentiated host tissue [14, 15]. Although sexually transmissible from founders to their progeny [8], these sequences exhibit considerable instability in adult somatic tissues and are progressively lost in parallel with ageing. Indeed, the reverse-transcribed structures are most abundant in young mice while being heavily reduced or totally absent in older individuals, as revealed by PCR analysis of DNA samples from mice at different ages [9].

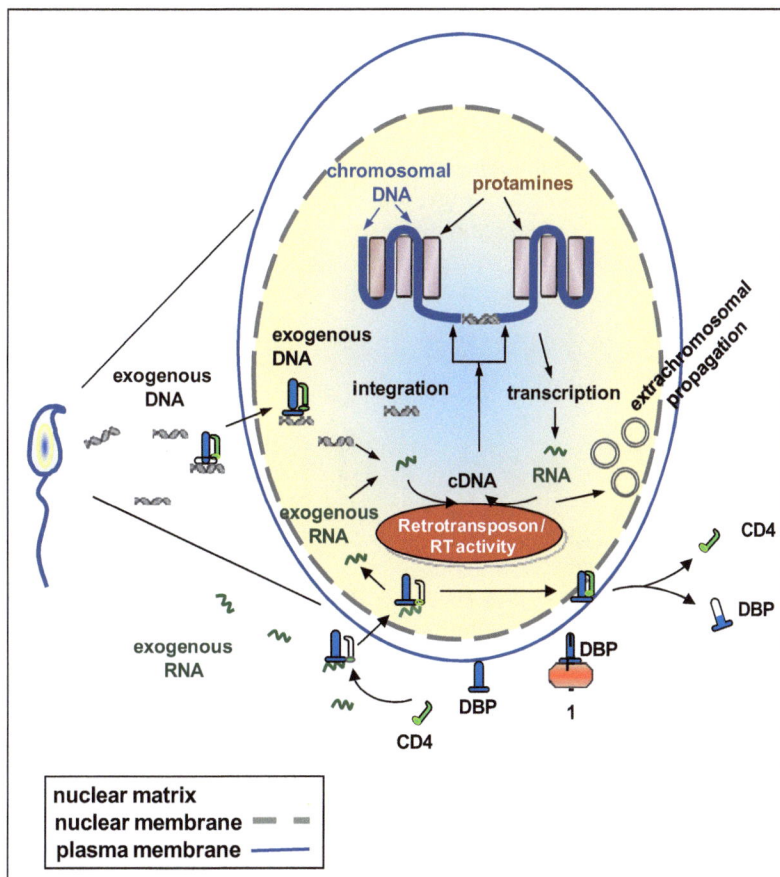

Figure 1: Model for the Genesis of Reverse-Transcribed Genetic Information in Sperm Cells. Exogenous RNA or DNA molecules bind to DNA-binding proteins (DBP, blue) localized on the cell membrane of intact spermatozoa, provided that the inhibitory factor (IF-1, red) does not interfere. The DNA/DBP complex interacts with CD4 molecules (green), is internalised in nuclei, and eventually dissociates at the level of the nuclear matrix, thus releasing the exogenous DNA; the DBP/CD4 protein complex can then be recycled to the cell surface (see Chapter 2). The internalised DNA molecules are transcribed in RNA and further reverse-transcribed in cDNA copies by the endogenous RT-containing retrotransposon machinery (matrix-associated brown oval). When exogenous RNA molecules are used,

they are also directly reverse-transcribed in cDNA, like the RNA transcribed from exogenous DNA molecules. A minor proportion of the resulting cDNAs eventually integrate into "accessible" sites of the sperm chromatin, which retain a nucleosomal organization and are probably located between adjacent protamine domains. The majority of cDNA copies, after delivery to the oocytes at fertilization, propagate instead as extrachromosomal structures (circles).

RETROGENES *VS.* TRANSGENES: THE ROLE OF THE PLASMA MEMBRANE

A key finding obtained in the laboratory of Ruyzo Yanagimachi was the combination of SMGT and ICSI, which yielded a substantial improvment in the production of transgenic animals with a radical change in the outcome compared to the original SMGT protocol [16; see also Chapter 2 of this book]. Central to their work was the use of demembranated spermatozoa incubated with foreign DNA (schematically represented in Fig. **2**), which were then microinjected into oocytes as a sperm/DNA complex following an ICSI procedure, due to the inability of demembranated spermatozoa to fertilise autonomously [reviewed by 17]. When the cell membrane barrier is bypassed, the exogenous DNA molecules (as opposed to the nucleic acid/protein receptor complexes forming in the original procedure, see Fig.1) have direct access to the perinuclear theca [reviewed by 18]; these molecules then integrate in the zygote genome *via* a DNA repair mechanism [19], yielding high rates of genuinely transgenic animals, with stably integrated transgenes that are inherited as Mendelian characters by the progeny. These features are clearly different from those observed in SMGT, in which exogenous sequences are taken up by specific factors and internalised as complexes in intact spermatozoa.

Putting aside for a moment the applications of the methods to the purpose of generating transgenics, some fundamental biological features elucidated by these results are worth stressing: a) exogenous nucleic acid molecules meet two possible fates when binding to sperm cells, and b) the plasma membrane is the key discriminating barrier between the two possible outcomes: when foreign sequences bind to intact sperm cells and are further internalised through a spontaneous process, they mainly generate extrachromosomal retrogenes, as illustrated in Fig. **1**; when exogenous sequences interact directly with the nuclei of demembranated spermatozoa (Fig. **2**), then regular transgenic animals are mostly obtained. Clearly these findings underscore the existence of a sophisticated machinery for processing foreign genetic information in sperm cells, which was almost serendipitously discovered in SMGT studies.

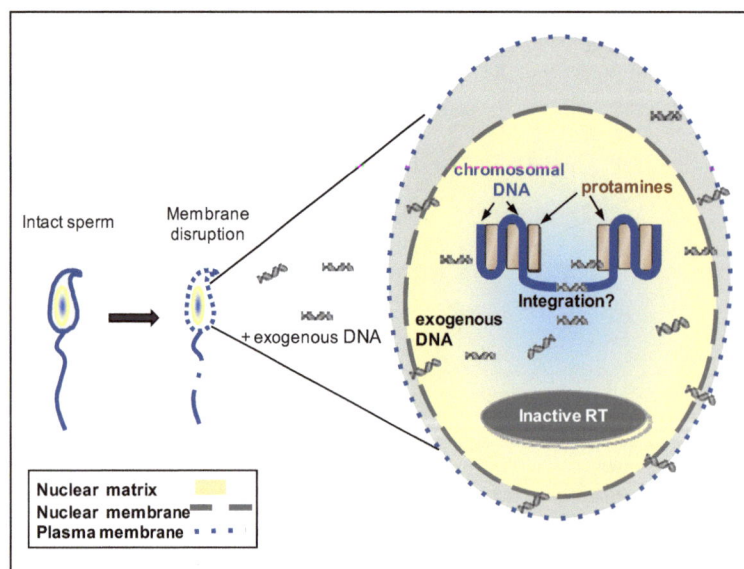

Figure 2: Schematic Representation of ICSI-Mediated Transgenesis using Mouse Sperm Cells. The plasma membrane is disrupted and membrane-disrupted sperm cells are incubated with exogenous DNA molecules. Under these conditions, naked DNA molecules have direct access to the sperm head. The sperm head/DNA complex is then microinjected into oocytes according to well-described ICSI protocols. Although the occurrence of integration events in the sperm genome cannot be ruled out, it is likely that most take place soon after fertilization in early zygotic or embryonic nuclei, as for microinjection.

THE EMERGENCE OF NON-MENDELIAN TRANSGENERATIONAL GENETICS

At this point, the RT-dependent mechanism operating in mature spermatozoa upon interaction with foreign DNA or RNA sequences emerges as a source of new genetic and phenotypic traits. The features of these traits render them clearly distinct from those encoded by chromosomal genes or transgenes, and reveal their predominant epigenetic origin. In other words, the apparently discrepant success of SMGT has in fact disclosed the emergence of novel non-Mendelian transgenerational genetics and RT-dependent inheritance, in which traits unlinked to chromosomal DNA sequences are transmitted to the progeny with a pattern that can involve germline transmission between generations.

Though poorly understood, phenomena of inheritance not explainable according to Mendel's laws are not new [reviewed by 20]. Indeed, this type of inheritance has been putatively correlated with a variety of late onset pathological syndromes in adult life, such as obesity, type-2 diabetes and cardiovascular diseases [reviewed by 21, 22], some tumor types [23], male reproductive defects [24] and schizophrenia [25]. One centrally puzzling aspect in these etiologically unrelated diseases is that no classical mutation or predisposing genetic lesion is identifiable in the genome of individuals that will be affected in adulthood, and yet inheritance through the germline operates from one generation to the next. This enigma may turn out to be explained, at least in part, by polygenic mechanisms. However, in the absence of the discovery of definitive polygenes underlying such disorders, such disorders present a challenge to conventional genetics, and raise the possibility that an additional as-yet-unidentified mode of genetic transmission may be involved.

RNA emerges as a potential player in this context. Indeed, RNA-mediated inheritance of epigenetic traits has recently been reported in mice [26, 27]: in those studies, suppression of the tyrosine kinase receptor *Kit* gene was lethal in homozygosis, whereas heterozygotes exhibited characteristic white spots in the tail tip and on feet. Unexpectedly, these features were also maintained in the genetically wild-type progeny. Moreover, that unexpected phenotype was associated with reduced *Kit* mRNA expression and accumulation of RNA of abnormal size in spermatozoa. That finding indicates that RNA can behave as an epigenetic modifier in phenotypic determination and can effectively replace or parallel DNA as the repository of genetic information. That result was interpreted as 'paramutation', a well-known phenomenon in plants that had never been described before in mammals [28]. The potential of RNA in substitution of DNA was further confirmed when the same group showed that cardiac hypertrophy can be induced and transmitted to the progeny by microinjecting in mouse zygotes either Cdk9 mRNA (a regulator of cardiac growth), or the related miR-1 RNA [27].

In our view, the RT-mediated process emerging from SMRGT shares striking analogies with RNA-mediated inheritance:

i.   Microinjection of RNA in zygotes (RNA-mediated) or loading intact spermatozoa with foreign DNA or RNA molecules (SMRGT RT-mediated), produce phenotypic effects in the offspring. In both cases, the exogenous sequences are delivered to embryos before pronuclei fusion.

ii.   The newly generated traits in both processes are transgenerationally inherited by the progeny in a non-mendelian fashion. The sexual transmission of these phenotypic traits implies that RNA, cDNA, or both types of molecules are stored in sperm cells.

iii.   The virtual absence of genes encoding new phenotypic traits is common to both systems. The mouse *Kit* model [26] and SMRGT-generated animals in our laboratory, which expressed reporter genes yet lacked any detectable lacZ [8] or EGFP [9] sequences when tissue DNAs were analyzed by Southern blot.

iv.   Newly induced traits are progressively lost with ageing.

v.   *Kit* RNA localizes in the subacrosomal segment of the mouse sperm head [26], *i.e.* the same site to which exogenous DNA or RNA molecules bind in SMGT [3, 7], which may identify an active subcellular compartment for RNA production and/or processing.

These facts underscore the key role played by RNA populations in both processes: in the case of RNA-mediated inheritance, the modifier RNA is thought to replicate through an RNA-dependent RNA polymerase (RdRP) [29], while a reverse transcriptase (*i.e.*, a RNA-dependent DNA polymerase activity) operates in the SMRGT process. We hypothesise that the RNA-dependent process, as for the RT-dependent mechanism, could also be in principle a source of new retrogenes. The study of Rassoulzadegan and coworkers shows that the "abnormal" RNA is specifically localized in spermatozoa: it may be speculated that this type of RNA, as well as the bulk population of RNAs that is physiologically stored in mature spermatozoa [reviewed by 30, 31], may be substrates for sperm RT. Reverse transcription of this and other RNAs may generate new retrogenes able to modify the phenotype and to be transmitted to the progeny in a non-Mendelian fashion.

Retroelements are well-known to respond to a variety of stimuli, including stress [32, 33]; their mobilization can affect the global expression not only of protein-coding genes [34, 35; also see references in reviews 36-38] but also of micro-RNA (miRNAs) and of ultraconserved non-coding elements (UCNEs) [39; also unpublished results of Sciamanna and Spadafora]. Extrachromosomal structures can also be generated in somatic cells in response to diverse stimuli [40, 41], and some contain retroposon-derived sequences [42, 43].

In retrospect, therefore, the variable outcomes of SMGT reflects the activation of the RT-dependent machinery in sperm cells; this activation is crucial in the response to foreign DNA or RNA binding and initiates a process that generates transcriptionally competent, non-integrated retrogenes [8, 9]. Seen in this perspective, SMGT is essentially an epigenetic phenomenon, the outcome of which may be regarded as a response of the stress-sensitive retrotransposon/endogenous retroviral machinery.

In terms of its potential application, SMGT can now be viewed as a particularly suitable method in those cases in which the genomic integrity of the recipient organisms must be preserved and the potential damage caused by integration events must be avoided. On these grounds, SMGT may offer a possibility for embryonic gene therapy based on the fact that both expression and transmission of novel phenotypic traits occur in the virtual absence of corresponding new genes.

CONCLUSIONS

Spermatozoa are a continuous source of surprising discoveries. These cells have turned out to do much more than traditionally thought based on their most obvious morphological and physiological features. Surprising was the discovery that the extremely compact structure of the sperm chromatin has room to accommodate ten of thousands of base pairs of exogenous DNA [44, 45]; even more surprisingly, these genetic sequences can be spontaneously delivered to oocytes at fertilization. In our view, these features are not merely the products of chance, but they have been selected and evolved to allow the binding, and penetration within the sperm nucleus, of factors that activate functions in the sperm chromatin; for example, activation of the process of sperm chromatin decondensation [46] comes to mind. These changes may be part of the cascade of events occurring during sperm capacitation [47] and may be regarded as 'preparatory' steps that sperm cells undergo prior to fertilising eggs. It can be hypothesized that the factors physiologically involved in these processes are released either by the oocytes themselves, or by the epithelium of the female genital tract, where spermatozoa cleared of the seminal fluid can take them up. The evidence recalled above indicates that one triggered function is the activation of an otherwise quiescent RT encoded by LINE-1 elements, which reverse-transcribes spermatozoal RNA substrate(s) into cDNA copies in zygotic pronuclei and in early preimplantation embryos, with the ensuing production of retrogenes. Importantly, new retrotransposition events have been demonstrated to occur in early embryos [48-50] and, furthermore, endogenous RT of LINE-1 origin is strictly necessary for preimplantation development [51, 52]. An equivalent mechanism may also be triggered in oocytes in response to sperm penetration, given that oocytes are potential sites of retrotransposition [53].

On these grounds, the generation of extrachromosomal structures in spermatozoa may be part of a retrotransposon-dependent mechanism which is activated in response to physiological stimuli of which fertilisation and embryo development offer conspicuous instances; the scope of the mechanism is probably the adaptation of the gamete genomes to the rapidly evolving growing embryo environment - possibly representing an extreme, albeit natural, type of adaptation. The mechanism would then be repressed at later developmental stages and in adult life but, as exemplified by stem cells niches [14, 15], it can be reactivated under particular conditions (*e.g.* stress, regeneration or others), to expand again the adaptive potential of the genome. When this occurs, novel epigenetic information can be produced which may turn out to be responsible for the genesis and non-Mendelian transgenerational propagation of many hitherto genetically enigmatic diseases.

REFERENCES

[1]   Lavitrano M, Camaioni A, Fazio V, *et al.* Sperm cells as vectors for introducing foreign DNA into eggs: genetic transformation of mice. Cell 1989; 57: 717-23.

[2]   Brinster RN, Sandgren EP, Behringer RR, *et al.* No simple solution for making transgenic mice. Cell 1989; 59: 239-41.

[3]   Spadafora C. Sperm cells and foreign DNA: a controversial relation. BioEssays 1998; 20: 955-64.

[4]   Smith KR. Sperm cell mediated transgenesis: a review. Anim Biotechnol 1999; 10: 1-13.

[5]   Smith K, Spadafora C. Sperm-mediated gene transfer: applications and implications. BioEssays 2005; 27: 551-62.

[6]   Maione B, Lavitrano M, Spadafora C, *et al.* Sperm mediated gene transfer in mice. Mol Reprod Dev 1998; 50: 406-09.

[7]   Giordano, R, Magnano AR, Zaccagnini G, *et al.* Reverse transcriptase activity in mature spermatozoa of mouse. J Cell Biol 2000; 148: 1107-13.

[8]   Sciamanna I, Barberi L, Martire A, *et al.* Sperm endogenous reverse transcriptase as mediator of new genetic information. Biochem.Bioph. Res Comm 2003; 312: 1039-46.

[9]   Pittoggi C, Beraldi R, Sciamanna I, *et al.* Generation of biologically active retro-genes upon interaction of mouse spermatozoa with exogenous DNA. Mol. Reprod Dev 2006; 73: 1239-46

[10]   Fuster CD, Farrell D, Stern FA, *et al.* RNA polymerase activity in bovine spermatozoa. J Cell Biol 1977; 74: 698-06.

[11]   Zoraqi G, Spadafora, C. Integration of foreign DNA sequences into mouse sperm genome. DNA & Cell Biol 1997; 16: 291-00.

[12]   Ward WS. Function of sperm chromatin structural elements in fertilization and development. Mol Hum Reprod 2009; 11.

[13]   Pittoggi C, Renzi L, Zaccagnini G, *et al.* A fraction of mouse sperm chromatin is organized in nucleosomal hypersensitive domains enriched in retroposon DNA. J Cell Sci 1999; 112: 3537-48.

[14]   Scadden DT. The stem-cell niche as an entity of action. Nature 2006; 441:1075-79.

[15]   Moore KA, Lemischka IR. Stem cells and their niches. Science 2006; 311:1880-85.

[16]   Perry AC, Wakayama T, Kishikawa H, *et al.* Mammalian transgenesis by intracytoplasmic sperm injection. Science 1999; 284: 1180-83.

[17]   Navarro J, Risco R, Toschi M, *et al.* gene therapy and intracytoplasmatic sperm injection (ICSI) – A Review. Placenta 2008; 29: 193-99.

[18]   Moisyadi S, Kaminski JM, Yanagimachi R. Use of intracytoplasmic sperm injection (ICSI) to generate transgenic animals. Comp Immun, Microbiol Infec Dis 2009; 32: 47-60.

[19]   Perry A. Hijacking oocyte DNA repair machinery in transgenesis? Mol Reprod Dev 2000; 56: 319-24.

[20]   Whitelaw NC, Whitelaw E. Transgenerational epigenetic inheritance in health and disease. Curr Opin Genet Dev 2008; 18: 273-79.

[21]   Gluckman PD, Hanson MA, Beedle AS Non genomic transgenerational inheritance of disease risk. BioEssays 2007; 29: 145-54.

[22]   Stoeger R. The thrifty epigenotype: an acquired and heritable predisposition for obesity and diabetes? BioEssays 2008; 30: 156-66.

[23]   Cheng RY, Hockman T, Crawford E, *et al.* Epigenetic and gene expression changes related to transgenerational carcinogenesis. Mol Carcinog 2004; 40: 1-11.

[24]   Anway MD, Cupp AS, Uzumcu M, *et al.* Epigenetic transgenerational actions of endocrine disruptors and male fertility. Science 2005; 308: 1466-69.

[25]   Lumey LH, Stein AD, Kahn HS, *et al.* Cohort profile: the Dutch hunger winter families study. Int J Epid 2007; 36: 1196-04.

[26] Rassoulzadegan M, Grandjean V, Gounon P, *et al.* RNA-mediated non-mendelian inheritance of an epigenetic change in the mouse. Nature 2006; 441: 469-74.

[27] Wagner KD, Wagner N, Ghanbarian H, *et al.* RNA induction and inheritance of epigenetic cardiac hypertrophy in the mouse. Dev Cell 2008; 14: 962-69.

[28] Chandler VL, Stam M Chromatin conversations: mechanisms and implications of paramutation. Nature Rev Genet 2004; 5: 532-44.

[29] Alleman M, Sidorenko L, McGinnis K, *et al.* An RNA-dependent RNA polymerase is required for paramutation in maize. Nature 2006; 442: 295-98.

[30] Miller D, Ostermeier GC, Krawetz SA. The controversy, potential and roles of spermatozoal RNA. Trends Mol Med 2005; 11: 156-63.

[31] Dadoune JP, Spermatozoal RNAs: what about their functions? Microsc Res Tech 2009; 72: 536-51.

[32] Rudin CM, Thompson CB. Transcriptional activation of short interspersed elements by DNA-damaging agents. Genes Chrom Cancer 2001;30: 64-71.

[33] Hagan RC, Rudin CM. Mobile genetic elements activation and genotoxic cancer therapy. Am J Pharmacogenomics 2002; 2: 25-35.

[34] Sciamanna I, Landriscina M, Pittoggi C, *et al.* Inhibition of endogenous reverse transcriptase antagonizes human tumor growth. Oncogene 2005; 24: 3923-31.

[35] Oricchio E, Sciamanna I, Beraldi R, *et al.* Distinct roles for LINE-1 and HERV-K retroelements in cell proliferation, differentiation and tumor progression. Oncogene 2007; 26: 4226-33.

[36] Lerat E, Semon M. Influence of the transposable element neighborhood on human gene expression in normal and tumor tissues. Gene 2007; 396: 303-311.

[37] Feschotte C. Transposable elements and the evolution of regulatory networks. Nat Rev Genet 2008; 9: 397-05.

[38] Goodier JL, Kazazian HH, Jr. Retrotransposons revisited: the restraint and rehabilitation of parasites. Cell 2008; 135: 23-35.

[39] Bejerano J, Pheasant M, Makunin I, *et al.* Ultraconserved elements in the human genome. Science 2004; 304:1321-25.

[40] Gaubatz JW. Extrachromosomal circular DNAs and genomic sequence plasticity in eukaryotic cells. Mutat Res 1990; 237: 271-92.

[41] Kuttler F, Mai S. Formation of non-random extrachromosomal elements during development, differentiation and oncogenesis. Sem Cancer 2007; 17: 56-64.

[42] Krolewski JJ, Rush MG. Some extrachromosomal circular DNAs containing the Alu family of dispersed repetitive sequences may be reverse transcripts. J Mol Biol 1984; 174: 31-40.

[43] Sherlina SS, Vladimirov VG. The involvement of different mobile LINE copies in blood plasma and extrachromosomal DNA of liver cells in systemic adaptive response. Ann NY Acad Sci 2008; 1137: 66-72.

[44] Lavitrano M, French D, Zani M, *et al.* The interaction between exogenous DNA and sperm cells. Mol Reprod Dev 1992; 31:161-69.

[45] Moreira PN, Perez-Crespo M, Ramirez MA, *et al.* Effect of transgene concentration, flanking matrix attachment regions, and recA-coating on the efficiency of mouse transgenesis mediated by intracytoplasmic sperm injection. Biol Reprod 2007; 76: 336-43.

[46] Caglar GS, Hammadeh M, Asimakopoulos B, *et al. In vivo* and *in vitro* decondensation of human sperm and assisted reproduction technologies. *In vivo* 2005; 19: 623-30.

[47] Baldi E, Luconi m, Bonaccorsi L, *et al.* Intracellular events and signalling pathways involved in sperm acquisition of fertilizing capacity and acrosome reaction. Front Biosci 2000; 5: E 110-23.

[48] van den Hurk JAJM, Meij IC, Seleme MC, *et al.* L1 retrotransposition can occur early in human embryonic development. Human Mol Gen 2007; 16: 1587-92.

[49] Garcia-Perez JL, Marchetto MCN, Muotri AR, *et al.* LINE-1 retrotransposition in human embryonic stem cells. Human Mol Gen, 2007;16: 1569-77.

[50] Kano H, Godoy I, Courtney C, *et al.* L1 retrotransposition occurs mainly in embryogenesis and creates somatic mosaicism. Genes & Dev 2009; 23: 1303-12.

[51] Pittoggi C, Sciamanna I, Mattei E, *et al.* A role of endogenous reverse transcriptase in murine early embryo development. Mol Reprod Dev 2003; 66: 225-36.

[52] Beraldi R, Pittoggi C, Sciamanna I, *et al.* Expression of LINE-1 retrotrasposons is essential for murine preimplantation development. Mol Reprod Dev 2006; 73: 279-87.

[53] Georgiou I, Noutsopoulos D, Dimitriadou E, *et al.* Retrotransposon RNA expression and evidence for retrotransposition events in human oocytes. Hum Mol Genet 2009; 18: 1221-28

CHAPTER 13

Conclusions

Kevin R. Smith[*]

Abertay University, United Kingdom

Abstract: Several experts in the various fields of SMGT have contributed valuable knowledge to this eBook. This chapter offers a synthesis of the key concepts covered throughout the book, and critically evaluates the current status and controversial aspects of SMGT.

Keywords: Augmented uptake, Autouptake, En masse GM, Episomal transgene, Gene targeting, Germline modification, Horizontal inheritance, Human artificial chromosome, SMGT scepticism, Zinc-finger nuclease.

Written by experts in their respective fields, the preceding chapters have described and critically analysed the various findings from SMGT experiments, with historical experiments, current work and future possibilities all being considered. Additionally, the fundamental science underpinning SMGT has been elucidated to as great an extent as is possible in this complex and emerging area of bioscience. These chapters therefore represent the most comprehensive summary assemblage of knowledge available to date in the field of SMGT. This final chapter provides a synopsis of the key aspects covered throughout the eBook, and identifies core principles and concepts central to the understanding of SMGT. In so doing, areas of controversy are highlighted and discussed.

A SYNTHESIS OF THE EXPERT CONTRIBUTIONS

Dating from the early 1970s, fundamental studies on spermatozoa biology met with the emerging technology of DNA modification to yield the first reports of SMGT. However, initial reports of successful experiments to generate transgenic animals *via* the incubation of sperm with naked DNA proved difficult or impossible to replicate, leading to a degree of skepticism and concomitant loss of interest in SMGT amongst many scientists in the transgenics field.

Aside from loss of interest resulting simply from the manifest empirical difficulties associated with the new technique of SMGT, a deeper concern was expressed by several biologists. This disquiet lay with the radical and disturbing possibility of 'horizontal' inheritance suggested by the initial SMGT experiments. In short, if sperm cells are easily able to pick up and transmit foreign DNA sequences, then considering that such molecules are abundant in nature, one would expect to see random appearances of novel sequences in animal genomes through the generations. Because inheritance is in fact observed to be an orderly process, *i.e.* following Mendelian rules, the conclusion of some biologists was that SMGT must be intrinsically impossible.

However, this deep skepticism towards SMGT proved difficult to sustain. While it should be expected that evolution has equipped sperm with robust mechanisms against potentially invading foreign DNA sequences, it does not necessarily follow that such protections are 100% perfect, or inviolable under all conditions. Thus, it is at least theoretically possible that mammalian genomes may occasionally pick up foreign sequences: indeed, recent research into the sequence structure of animal genomes has revealed the presence of various foreign sequences, a finding that is compatible with the possibility of 'natural' SMGT. Against this, it may be argued that entry of such foreign sequences into the sperm genome might have been facilitated by special factors, such as viral infections (of spermatozoa or spermatogonia), or unusual biochemical imbalances within the reproductive tract. However, as described at various points in this

*Address correspondence to Kevin R. Smith: School of Contemporary Sciences, Abertay University, Dundee, DD1 1HG, United Kingdom; Tel: 44 (0)1382 308664; E-mail: k.smith@tay.ac.uk

eBook, laboratory methods of SMGT (other than naked DNA incubation) have been developed which in a sense mimic nature by facilitating transgene entry into sperm. There is no reason to assume that sperm cannot be induced to take up exogenous DNA molecules, given appropriate conditions.

Subsequently, some of the most important research into SMGT has focused on the elucidation of parameters and underlying mechanisms associated with SMGT. From such work, some unique and indeed surprising insights into sperm biology have been obtained, including the finding that sperm -hitherto considered to be one of the most inert cell types at the nuclear level-appear to be endowed with endogenous reverse-transcriptase activity. Such findings, although themselves relatively preliminary and requiring of further explication, have enabled the building of intriguing putative models of transgene uptake and transmission by sperm. These models, which were considered in depth in Chapter 12, entail a complex interplay between sperm nuclear processes and transgene molecules, and represent a potential paradigm shift in our understanding of sperm-mediated inheritance.

Claims of successful SMGT have continued to be published since the landmark (and most controversial) work of the late 20th century. Despite such publications, and despite the inherent attractiveness of SMGT as a low cost, *en masse* means of animal genetic modification, SMGT has not yet become a standard, reliable technique of transgenesis. In part, this apparently paradoxical situation is due to the lack of 'true' transgenic animals generated by the method - *i.e.* animals that contain integrated transgene sequences and are able to pass their genetic modifications to subsequent generations in a normal Mendelian manner.

Research evidence, assimilated gradually over recent years, now suggests that the lack of genetic transmission in such cases is due to the generation of episomal structures following uptake of transgenes by sperm. Such transgene episomes may exist within the cells of the founder (first generation) animals, and can give rise to phenotypic manifestations. However, such animals generally exhibit high levels of transgenic mosaicism, and fail to reliably transmit the transgene when reproducing. Specifically, it appears to be the interaction between the sperm membrane and transgene DNA that leads to this anomalous pattern of transgene maintenance and inheritance. By contrast, delivery methods that bypass the sperm membrane tend to be associated with transgene integration into the host genome, thus yielding true transgenic animals. These findings suggest that, for applications requiring true transgenics, simple incubation of sperm with naked DNA will not suffice; instead, alternative methods that permit transgenes to enter the sperm nucleus without direct membrane interaction look more promising. Of the several available alternative SMGT methods, SMGT combined with ICSI (known as transgenICSI or ICSI-TR) offers the clearest form of membrane bypass. It is thus unsurprising that, amongst the various forms of SMGT, transgenICSI has proven, in a number of (mostly murine) systematic experiments, to be the most effective method for the generation of true transgenic animals.

TransgenICSI possesses a key advantage over the standard means to producing transgenics, *i.e.* pronuclear microinjection (PMI): it is better able to deliver extremely large transgene constructs, including artificial chromosomes, such as YACs, MACs and HACs. Amongst several benefits, such large transgene constructs permit more faithful expression of transgenes. TransgenICSI also offers lower frequencies of mosaicism than does PMI. However, in comparison to other forms of SMGT, transgenICSI requires micromanipulation of embryos: this necessity removes the *en masse* advantage of SMGT, and increases its costs. Thus, other SMGT methods that permit membrane bypass at least to some extent, while not necessarily allowing the use of very large transgene constructs, remain attractive. Examples of such SMGT 'augmentation' variants include liposome-mediated gene transfer, electroporation, viral vectors, linker-mediated uptake and restriction enzyme-mediated integration (REMI). Most recently, nanobiotechnological tools have been applied in the context of SMGT. Such methods, most of which are at relatively early stages of development, include nanocomposite-based vesicles, nanoparticle bombardment, nanotubes, and magnetofection.

Alternatively, in a variant of SMGT termed testis-mediated gene transfer (TMGT), transgenes may be injected directly into the testis, thus leading to the *in vivo* production of sperm containing integrated transgene sequences. In recent years a number of TMGT experiments, involving a range of animal types, have been concluded successfully. Thus, TMGT may well prove to be an important future form of animal genetic modification.

Successful preliminary work has been published (and is discussed at various points in this eBook) on all of the foregoing SMGT variants, suggesting that at least some of these methods are able to produce true transgenics. These variant forms of SMGT have been first applied in the murine context. The laboratory mouse is of course well reputed for its usefulness in biology, on account of various factors including low generation time, low cost, and availability of defined genetic strains. However, SMGT *per se* is also applicable to an extremely wide range of other animal types and species, including large domestic animals. In the context of biotechnology and medicine, SMGT methods that are able to generate true transgenics for these animal types clearly offer significant future potential. Such SMGT-generated transgenic animals have the potential to be used to express valuable pharmaceutical products, produce organs for xenotransplantation, and serve as animal models of human disorders. The major animal types that have yielded positive results in the context of SMGT are cattle, pigs, and chickens. Time will tell which (if any) of the variant forms of SMGT will prove to be of sufficient practical worth to the extent that they come to match or supplant current methods of transgenesis such as PMI.

One promising branch of biotechnology involves the production of transgenic animals for food purposes. Goals include the generation of animals that grow more rapidly, convert their feed more efficiently, resist diseases, and produce products of higher nutritional value. In addition to land animals, such as cattle, pigs and chickens, aquatic species are of great interest in this regard. Given that fish employ external fertilisation, SMGT involving simple incubation of fish sperm with transgene DNA initially appeared very attractive: however, the previously mentioned issue of episomal transgene maintenance and concomitant problems with expression and inheritance would seem to preclude this approach as a means to producing inheritable genetic modifications in fish and other aquatic animals. By contrast, preliminary experiments indicate that testis-mediated forms of SMGT may be very effective for aquatic species.

If SMGT, in any of its forms, is to be developed systematically, is essential that the underlying mechanisms be fully elucidated. Based on a limited but growing set of fundamental data on the sperm nuclear environment, and incorporating the above-mentioned empirical findings from SMGT experiments, certain models for the uptake and transmission of transgenes in SMGT have been proposed. The current models differentiate between two transgene uptake routes: [a] following interaction between the intact sperm membrane and the transgene molecules, *via* sperm-DNA incubation protocols; and [b] following transgene bypass of the sperm membrane, as in augmented SMGT approaches, typified by transgenICSI. In the former case, the best model currently available proposes that the exogenous DNA is transcribed into RNA which is then reverse transcribed into an episomal cDNA sequence, which may be propagated as an episomal structure, maintained in a highly mosaic pattern in the cells of the founder, and transmitted in a non-Mendelian manner with concomitant reduction in copy number and fadeout of phenotypic effects in the subsequent generation(s). By this model, transgene integration is envisioned as a rare occurrence; thus, few true transgenic animals are expected from simple sperm-DNA incubation protocols. By contrast, to explain cases where the sperm membrane has been bypassed, the best model we have proposes that transgene molecules integrate into the sperm genome, *via* nonhomologous recombination, to generate a high proportion of genuinely transgenic animals. More research is needed to further refine these groundbreaking models.

CONTROVERSIAL ISSUES

The SMGT field was kick-started upon publication of the well-known landmark *Cell* paper in 1989, claiming that true transgenic mice had been obtained following a simple sperm-DNA incubation protocol [1]. However, these potentially revolutionary results could not be replicated, despite immediate concerted attempts to do so by several independent groups around the world [2]. This situation understandably led to substantial skepticism towards SMGT.

Accumulated evidence, coupled with theory, now suggests that intact sperm are unable to efficiently transmit naked DNA molecules as integrated transgenes. However, the notion of simple sperm-DNA incubation, otherwise known as 'autouptake', as a means for genetic modification has not gone away. There are two main reasons for this. Firstly, the ability to easily induce sperm to act as vectors, simply through naked DNA incubation, would have very far-reaching implications for the understanding of inheritance, as well as for biotechnology and medicine. Thus, because this form of SMGT would be of great value, it would be premature to close one's mind to the possibility that it might somehow be achievable. The second reason why this issue is still live is that reports of true transgenics being generated by autouptake protocols have been published subsequent to 1989, with such reports continuing to appear until the present time - as is reflected in various reference citations throughout the chapters of this eBook.

The effectiveness of the transgenICSI approach demonstrates beyond reasonable doubt that sperm can act as transgene vectors. However, because this form of SMGT requires microinjection of the sperm head into individual oocytes, transgenICSI does not constitute an *en masse* tool of genetic modification. Similarly, the effective methodology of TMGT, although generally less onerous than transgenICSI, does not constitute a truly *en masse* method.

Yet it is the *en masse* nature of SMGT that makes it so potentially attractive, in terms of permitting easy, rapid and inexpensive genetic modification. Indeed, should SMGT based on autouptake followed by artificial insemination ever be achievable as a means to reliably insert transgenes into the host genome, it would represent an enormous quantum leap forward in the field of genetic engineering. Amongst other benefits, such an advance would permit the rapid genetic improvement of agricultural animals, and an acceleration of the development of animal bioreactors and multi-transgene animals for xenotransplantation. More futuristically, it could open the door to widespread germline improvement within human populations, for example by permitting the addition of tumour suppressor genes, or anti-HIV genes, into the human germline, in a fashion somewhat akin to that of present-day immunisation programs.

How can the reportedly successful results from simple autouptake protocols be explained? Firstly, a distinction is necessary: on the one hand, *in vitro* experiments frequently demonstrate the ability of spermatozoa to bind naked DNA molecules; on the other hand, the generation through autouptake of true transgenics -meaning animals whose cells indubitably contain transgene sequences in integrated form and are able to transmit such sequences to the next generation(s) in a Mendelian fashion- is a much less frequent occurrence. It is the latter claimed results that are controversial, not the former.

One possibility is that the presently accepted theory, as formalised in the above-mentioned SMGT models, is erroneous. However, the current SMGT models appear to be fully compatible with the majority of published data. By contrast, it is notable that the majority of the continuing claims of success for true transgenic animal generation through autouptake have been made by a limited number of researchers. It is not the case that a wide range of independent researchers are obtaining similar results. To take a broad perspective, scientific validity is dependent upon the ability to replicate experiments. This is especially true in two contexts: [a] where the claimed results conflict with orthodox explanations; and [b] where the findings do not represent absolutes, but rather come in the form of a claimed 'signal' amidst background 'noise'.

Claims that true transgenics can be produced by simple autouptake conflicts with orthodox science, because such claims appear to violate the well-founded observation that genetic inheritance is orderly, and 'horizontal' inheritance of foreign sequences is very rare. It is a central tenet of scientific epistemology that extraordinary claims require extraordinary evidence. Considering the fact that claims for successful sperm-DNA incubation run counter to established rules of inheritance, empirical evidence from such experiments would have to be particularly strong if they were to be accepted at face value.

On first perusal, transgenic animals, as an experimental outcome, would not appear to represent anything other than an absolute form of data. However, this is something of an illusion. It is clear from most of the positive reports that the presence of transgenes is usually detected by PCR analysis. The power of PCR is such that it is not too difficult to accidentally generate false positive data, for example by DNA contamination. Moreover, there are several ways in which other data errors can occur, for example in terms of problems with record keeping in animal houses. In summary, it can be argued that, if the claims that simple autouptake SMGT protocols can produce real transgenics are to be accepted, a much more robust body of data will need to be assimilated, crucially involving replication by independent research groups.

If it is correct that there are no good reasons to believe that sperm can be made to act as transgenic vectors *via* simple DNA incubation, it brings to the fore the following question: should the positive findings from the SMGT augmentation methods (such as liposomes, nanoparticles, magnetofection, etc) be accepted at face value? As an intermediate possibility between effective transgenICSI and the least plausible DNA autouptake approaches, SMGT augmentation techniques are of great interest. Most of these techniques can be used *en masse,* and in theory may succeed in producing genuinely transgenic animals, insofar as any such technique allows the transgenes to bypass the sperm membrane. However, in contrast to transgenICSI, these augmentation techniques do not constitute a complete avoidance of membrane interaction. It is clear that, in any cell type (including sperm), liposome-mediated DNA delivery requires complex interactions with the outer membrane of the cell and also (at least in some cases) with endosomic membranes. Similarly, electroporation proceeds either by the generation of minipores or *via* pre-existing channels in the membrane: in either case, membrane interactions are involved. Other augmentation methods, as described in this eBook, are also likely to unavoidably entail some membrane interaction. However, it is simply not known to what extent such interactions (and which particular interactions) may trigger the episomal pathway. Certainly, the number of true transgenics -able to transmit the transgene to the next generation- reported to have been produced by the various augmentation methods has been disappointingly low, which is perhaps suggestive of the episomal mode being dominant in this context. However, there are no good reasons to assume that augmentation techniques will always, by necessity, lead to the episomal pathway being triggered. Unfortunately, the limited extent of the available data means that it is presently impossible to know to what extent the successful reports represent mere 'noise' or genuine 'signal'. Clearly, it is highly desirable that much more research be conducted into the various augmentation approaches described in this eBook.

It is important to note that, even where true transgenics are not obtained, the episomal structures generated by direct sperm-DNA interaction remain of great interest, for several reasons: (a) these structures can be transcriptionally competent; (b) they can be sexually transmitted (albeit in a non-Mendelian fashion); and (c) they are able to generate new phenotypic traits in adult animals. These features suggest possible future applications of SMGT in its autouptake guise, in the contexts of animal biotechnology and gene therapy.

In the vast majority of SMGT experiments that report the generation of true transgenic animals, genomic integration is random. Random integration, as opposed to gene targeting, represents a conceptual limitation associated with SMGT. PMI has also been limited by the same restriction, thus placing both approaches on a conceptual par with regards to the mode of transgene integration. However, very recent work now suggests that co-injection of custom-made zinc-finger nuclease (ZFN) molecules may enable gene targeting to be achieved *via* PMI [3]. ZFNs arguably represent a quantum leap for targeted genetic engineering; however it is unclear whether such molecules could be combined with SMGT protocols in order to enable sperm to act as gene targeting vectors. If this does prove possible, then SMGT will surely gain greatly increased prominence. Conversely, should sperm vectors and ZFNs prove incompatible, SMGT may lose a significant amount of its attractiveness. This is an area that requires intensive research.

CONCLUDING REMARKS

SMGT is a remarkable area in modern biology. The range of potential and actual modes of SMGT are now very wide, ranging from simple autouptake methods, to augmented techniques, and to transgenICSI and TMGT approaches. In principle, SMGT offers radical improvements in our ability to engineer animal -and even human- genomes. The fact that SMGT to date remains a relatively underdeveloped science may be a consequence of the initial disappointment and resulting skepticism generated two decades ago when apparently promising landmark work failed to be replicable. To some extent the controversial nature of SMGT continues to the present day, but there can be little doubt that SMGT, at least in some of its guises, offers great potential for biology, biotechnology and medicine.

REFERENCES

[1] Lavitrano M, Camaioni A, Fazio VM, *et al.* Sperm cells as vectors for introducing foreign DNA into eggs - genetic-transformation of mice. Cell. 1989; 57: 717-23.

[2] Brinster RL, Sandgren EP, Behringer RR, Palmiter RD. No simple solution for making transgenic mice. Cell 1989; 59 :239-41.

[3] Meyer M, de Angelis MH, Wurst W, Kuhn R. Gene targeting by homologous recombination in mouse zygotes mediated by zinc-finger nucleases. Proc Natl Acad Sci USA 2010; 107: 15022-26.

INDEX

www.ingramcontent.com/pod-product-compliance
Lightning Source LLC
Chambersburg PA
CBHW041715210326
41598CB00007B/665

* 9 7 8 1 6 0 8 0 5 4 3 2 9 *